TECHNIQUES FOR SURVIVING THE MOBILE DATA EXPLOSION

TECHNIQUES FOR SURVIVING THE MOBILE DATA EXPLOSION

DINESH CHANDRA VERMA
PARIDHI VERMA

IEEE PRESS

WILEY

Library of Congress Cataloging-in-Publication Data:

Verma, Dinesh C.
   Techniques for surviving the mobile data explosion / Dinesh Chandra Verma, Paridhi Verma.
      pages   cm
   Includes bibliographical references.
   ISBN 978-1-118-29057-6 (pbk.)
1. Mobile computing.   I. Verma, Paridhi.   II. Title.
   QA76.59.V47 2014
   004.167–dc23

                                                          2013033685

ISBN: 9781118290576

10 9 8 7 6 5 4 3 2

Dedicated to our wonderful children Archit and Riya.
Thanks to our parents for their love, endless support,
and encouragement.

CONTENTS

PREFACE

Mobile devices have transformed the way we conduct business and lead our lives. Not only are we able to have a phone conversation with others using a Smartphone, we are able to check our e-mail, read a book, make a purchase, book airline tickets, snap a picture, connect with old friends, and play games using a device that we can carry with us anytime anywhere. The resulting transformations in society and our personal lives are apparent to all.

Despite all the wonderful capabilities offered by mobile devices, there is a significant hurdle that can easily turn into a showstopper preventing us from fully attaining the benefits possible from them. The amount of bandwidth available for communication using mobile devices is limited. In one part of the network, namely, the air interface, the bandwidth limitations are due to physical laws and government regulations. In other portions of the network, for example, the cellular backhaul, the bandwidth limitations are due to business reasons and the complex ecosystem of mobile data and applications. As mobile applications continue to become more popular, approaches to deal with the resulting mobile data growth will be needed by operators of mobile networks, developers of mobile applications, and enterprises, who are all increasingly relying on mobile applications.

Mobile data growth is a problem that many people are aware of, but it is hard to find many solutions that address that problem. The problem is especially hard to solve because it spans three complex and sophisticated technical fields, the field of cellular networks, the field of TCP/IP-based data networks, and the field of mobile application development. Addressing the challenges of mobile data growth requires looking across all three fields, which is not always easy for experts in one individual area. This book tries to span these three fields and provides sufficient details for an expert in one field to address the challenges involved in the other two fields.

This book is about the different approaches that can be used to address the challenges of limited bandwidth available to mobile applications by different members of the mobile data ecosystem. The mobile data ecosystem consists of many different constituencies, including the mobile network operators, the developer of mobile applications, and the enterprises that deploy mobile applications for their employees. This book takes the perspectives of each of the different constituencies in the mobile data ecosystem, examines the challenges they will encounter due to the growth of mobile data, and enumerates the various approaches that are available to them to address those challenges.

WHO WILL BENEFIT FROM THIS BOOK?

This book is intended for executives, technology leaders, graduate students, network practitioners, application developers, and system architects who are interested in mobile data or mobile applications. If you are an executive or technology leader for a mobile network operator, a company that develops a mobile application, or an enterprise that uses mobile applications, this book will be useful for you to get an understanding of the challenges associated with mobile data communications. This book provides a holistic overview of the challenges associated with mobile data communication. The description is technical yet generic enough to provide a good summary of the different options available for dealing with mobile data explosion.

If you are a network engineer, network architect, or network strategist working for a mobile network operator, you will find this book to be of interest. It will provide you a summary of the issues involved when Internet technology intersects with cellular network technology and outline the approaches that are available to address them. The book discusses various approaches that can be used to maximize the amount of bandwidth that can be put through an existing network infrastructure, methods that reduce the cost of operation of the network infrastructure, and some new ways in which the data flowing through the network can be monetized.

If you are a software architect or software developer involved in writing mobile applications, you will find this book to be of interest. You will find an overview of techniques you can use to make your application run better in the presence of constrained resources on the mobile device, including approaches to minimize power consumption and approaches to reduce network bandwidth consumption. You will also find a set of best practices that are compiled from a variety of sources in the technical community.

If you work for a company that makes equipment for mobile network operators, you will find this book to be of interest. This book will provide you with an overview of Internet-based data communications and issues involved in mobile application development. This will also give you a broad overview of techniques you can implement in your products to obtain better bandwidth efficiency and create new services for your customers.

If you are a systems architect or software architect working for an enterprise that is using mobile applications, you will find this book to be of interest. In addition to learning about the ways in which you can make mobile applications power-efficient and bandwidth-efficient, you will find a discussion of the various issues associated with the use of mobile applications in the enterprise. You will also find a summary of various approaches to deal with mobile data growth in networks, a subset of which may be beneficial for you within the enterprise network infrastructure.

If you are a graduate student or academic working in the field of mobile computing or mobile networking, you will find the overview of various techniques that lie at the intersection of cellular networks, IP networks, and mobile applications to be of relevance and interest.

Finally, if you are a practitioner in one of the three technical fields of cellular networks, TCP/IP-based data networks, or mobile application development and are interested in learning about the other two technical fields, you will find this book to be of interest. This book will give you a broad overview of the other two areas and discuss how approaches that cut across all the three different areas can help in addressing the challenge of mobile data growth.

WHO IS THIS BOOK NOT FOR?

This book covers a broad overview of three distinct technical fields, which also means it does not go very deep into any single technical field. If you are looking for a detailed overview of one of the technical fields, this book is not for you. The book covers only a broad perspective of each field, which is just sufficient to understand the challenges of mobile data growth and approaches to address them.

If you are looking for techniques to develop mobile applications using a specific operating system or specific device, this book is not for you. This book provides an overview of approaches to write efficient applications in a general way and helps you understand the technical foundations behind efficient application development. However, it will not provide you a detailed recipe for applications in any specific environment.

If you are looking for bandwidth optimization products or technology offerings from a specific network equipment manufacturer or any other company, you will not find this book of interest. This book will help you understand the general technical principles that drive various bandwidth optimization products, but it does not examine the products of any specific company.

ORGANIZATION OF THE BOOK

The material in this book is organized into 14 chapters, which can be roughly divided into three sections. The first section consisting of the first four chapters provides an overview of the three different technical fields that are involved in the challenge of mobile data growth, an overview of the mobile data ecosystem, and a general overview of the techniques for bandwidth optimization and cost reduction. The second section can be viewed as a section geared towards mobile network operators and consists of the next five chapters. It examines different aspects related to bandwidth optimization and services that can help a mobile network operator monetize mobile data. The third section consisting of the next four chapters can be viewed as a section for mobile application developers and enterprise users. These chapters consider approaches to write efficient applications and address issues with mobile application deployment in the enterprise. The final chapter in this book discusses some topics that are relevant to, but not directly related to, the main subject of this book.

Chapter 1 provides a summary of the three technical fields that need to be considered to understand mobile data growth. These three fields are cellular networks, TCP/IP data networks, and mobile application development. This chapter provides a high-level overview of the three technical areas with a brief explanation of their salient features.

Chapter 2 provides an overview of the mobile data ecosystem, describes the different constituencies in the mobile data ecosystem, and discusses the nature of mobile data growth and the impact of that growth on each of the constituents of the ecosystem.

Chapter 3 examines the various approaches that can be used within the network to manage the bandwidth overload problem. The problem is presented in the context of an idealized network environment, and different solutions are presented in the context of that environment. This chapter discusses a variety of network optimization methods, which will reduce the amount of data that needs to be sent over a bottleneck link.

Chapter 4 provides an overview of approaches that can be used within a network to reduce its cost of operation. Techniques discussed in this chapter include infrastructure sharing, application of consolidation technologies like Cloud, and concepts like network function virtualization.

Chapter 5 combines ideas from Chapters 3 and 4 and applies them to the specific case of radio access network. Both technical and nontechnical approaches that can be applicable to radio access networks are considered. Technical approaches include upgrading the network infrastructure and augmenting the network bandwidth, traffic offload, rate control, and service differentiation. Nontechnical approaches include different pricing plans, incentives against bandwidth hogs, and prompting users to switch over to Wi-Fi networks when appropriate.

Chapter 6 combines ideas from Chapters 3 and 4 and applies them to the specific case of radio backhaul network, one portion of the network operated by a mobile network operator. The chapter looks at both technical approaches as well as nontechnical approaches to reduce cost of operations and reduce the bandwidth transferred on this part of the network.

The next three chapters discuss some of the new services that can be used by a mobile network operator to obtain more money from data flowing over its networks. Chapter 7 focuses on new data-oriented services that a mobile network operator can offer to its current subscribers, while Chapter 8 provides an overview of some services that a mobile network operator can offer to enterprises or its business customers. Chapter 9 discusses some of the new services a mobile network operator can offer to application service providers that provide over-the-top services. Although application service providers and mobile network operators often have a competitive relationship, a variety of data-oriented services can be offered by a mobile network operator to application service providers benefiting both of them.

Chapter 10 provides an overview of mobile applications and discusses some of the challenges associated with the development of mobile applications. Some of these challenges are standard software development challenges associated with developing software on a diverse number of platforms, while other challenges are unique to a mobile development environment.

Chapter 11 looks at the issue of power efficiency in mobile applications. Starting from a simple model consisting of three abstractions of resources, resource managers and resource consumers, the chapter examines how resource consumers can operate so as to reduce power consumption of the different resources. The chapter also provides a set of best practices that can help in writing power-efficient applications.

Chapter 12 explores the issue of developing bandwidth-efficient mobile applications. It presents a set of techniques that can be used to make applications more bandwidth efficient and a set of best practices that can help in writing bandwidth-efficient applications.

Chapter 13 examines the issues of mobile device growth and adoption among enterprises. Enterprises face challenges associated with data security, enabling mobile access to existing applications, and mobile device management. This chapter discusses some of those challenges and the various techniques available to the enterprise to address those challenges.

Chapter 14, the final chapter in the book, looks at some of the topics that are related to the issue of mobile data growth, but not directly associated with it. These topics include Machine-2-Machine communications, the Internet of Things, the transformation in business processes that can be brought about using mobile applications, and applications such as participatory sensing.

ABOUT THE AUTHORS

Dinesh Chandra Verma, Fellow IEEE, is an IBM Fellow and Department Group Manager of the IT & Wireless Convergence area at IBM T J Watson Research Center, Hawthorne, New York. He received his doctorate in Computer Networking from the University of California Berkeley in 1992, bachelors in Computer Science from the Indian Institute of Technology, Kanpur, India, in 1987, and Masters in Management of Technology from Polytechnic University, Brooklyn, NY, in 1998. He has served in various program committees, IEEE technical committees, and editorial boards and has managed international multi-institutional government programs. His research interests include topics in wireless networks, network management, distributed computing, and autonomic systems. He holds over 50 U.S. patents related to computer networks and has authored over 100 papers and 9 books in the area.

Paridhi Verma is a marketing manager at IBM. She has 10 years of experience working at IBM Research as a network and security software engineer where she designed and implemented secure electronic commerce systems. She has several years of experience in developing strategic messaging and client value proposition for different industries working as a marketing and communications manager. She has a Master of Science in Electrical Engineering from NYU Poly and holds a patent and a publication for the Internet emergency alert system she designed while at research. She has also authored several books for children, which can be found at chandabooks.com.

SECTION 1

INTRODUCTION AND GENERAL OBSERVATIONS

CHAPTER 1

TECHNOLOGIES SUPPORTING MOBILE DATA

1.1 INTRODUCTION

The popularity of mobile devices is growing exponentially, and the number of such devices deployed worldwide is rising rapidly. These devices come in many forms such as Smartphones, Personal Digital Assistants, and tablets. Another important category of mobile devices are laptops where the use of network interfaces that support mobile wireless data, either via cellular data networks or through Wi-Fi, is commonplace.

The increase in number of mobile devices has led to the development of many new applications targeted at both businesses and consumers. The count of such applications runs into several hundreds of thousands for any of the popular mobile device platforms. Some of these applications require very little network data exchange but some, for example, applications which allow a user to watch streaming video on the mobile device can use up a tremendous amount of data in a very short amount of time.

As a result, the amount of data that is sent over the mobile cellular networks keeps on increasing steadily. The amount of data growth in mobile networks is tracked by various organizations. Studies conducted by several companies [1,2] indicate that mobile data growth has approximately been tripling every year since 2009. A significant growth in the data came from the

Techniques for Surviving the Mobile Data Explosion, First Edition.
Dinesh Chandra Verma and Paridhi Verma.
© 2014 The Institute of Electrical and Electronics Engineers, Inc. Published 2014 by John Wiley & Sons, Inc.

transmission of video on the network, which accounts for more than half of the total mobile data. Furthermore, predictions made by both these studies, as well as other studies, show no signs of slowing down of this trend of mobile data growth.

The bandwidth demands from all of the mobile applications that exist currently and will be developed in the future will likely exceed the network capacity that can be offered by the currently deployed wireless cellular networks. This mismatch in capacity can be handled in a variety of ways, and the different techniques that can be used to solve the capacity mismatch are the subject of this book.

In an ideal world, one would simply upgrade the network infrastructure so that all of the challenges associated with limited bandwidth disappeared. In real life, this simple solution comes with an enormous price tag. There are several constituencies with some vested interest in mobile data communications, for example, mobile device users, mobile network operators, mobile application developers, enterprises using mobile computing applications. Each of these constituencies would like the bulk of cost required for bandwidth upgrade to be borne by some other constituency. The actions to address the capacity mismatch that are within the control of each constituency are also different. An introduction to the various constituencies in the mobile ecosystem and the implications of mobile data growth on each of the constituencies is provided in Chapter 2 of this book.

Approaches that can be used to address mobile data growth have technical complexities in addition to the ecosystem complexities. Mobile data communications are at the intersection of three different, though related, technical fields, namely (i) mobile applications, (ii) Internet, and (iii) cellular networks. Approaches to address mobile data growth need to span three technical areas, which is more complex than any approach that could be addressed within one single technical area.

The relationship and interaction between these three areas can be understood by a reference to Figure 1.1. The figure shows a high-level layout

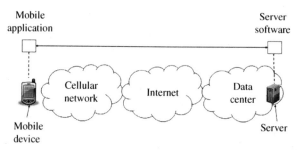

FIGURE 1.1   Mobile applications data communication infrastructure.

of the communication infrastructure needed for a mobile application that is exchanging data with another computer on the Internet.

Mobile applications are software components that run on mobile devices such as a Smartphone or tablet computer. They exchange data with application components running on servers, shown as server software in Figure 1.1. In order to communicate with each other, the mobile application and server software use a set of conventions called a communication protocol. The Hypertext Transport Protocol (HTTP) is an example of a common protocol used in this manner. The information exchanged by this protocol traverses a number of networks between the mobile device and the server. These networks include a cellular network that connects the mobile device wirelessly to the Internet, and the Internet that provides a means to connect the cellular network to the data center where the server may be physically located.

As the figure demonstrates, mobile applications run using communication protocols overlaid on top of Internet protocols (IPs) that are overlaid partially on top of cellular networks. Each of these three technical fields has its own sophisticated set of well-developed technologies and best practices. Two of these technical fields, namely cellular networks and Internet are specific cases of computer communication networks and have some terminology and design principles in common. However, due to historical reasons, the Internet and the cellular networks have each evolved terminology and mechanisms that are very different. Managing the growth of data due to mobile applications requires an approach that can span across each of these three technical fields and take into account the idiosyncrasies and characteristics of each of these fields.

In this introductory chapter, we provide a high level overview of these technology areas, beginning with an overview of data networks, followed by a discussion about the Internet (an instance of a data network) and the cellular networks (another instance of a data network), and the mobile applications protocols.

1.2 COMPUTER COMMUNICATION NETWORKS

The basic function of a computer communication network is to enable two or more computers to exchange data with each other. To enable this exchange, the computers need to agree to a set of conventions that they can all understand and agree upon. Such a set of conventions is called a communication protocol. Computer communication networks consist of a set of communication protocols that build upon each other in a layered manner.

In order for two computers that may be in two different continents to communicate, several layers of communication protocols are needed. Let us consider a simple interaction, one that happens when a user types in the address of a website like http://www.chandabooks.com in a web browser on his Smartphone, and another computer sends back some data which is displayed to the user. In order to facilitate this exchange, a protocol called HTTP is used between the Smartphone and the IEEE web server. The protocol defines the formats in which the requests from the Smartphone to the web server ought to be sent, and how the web server will respond back to it. This protocol is built using the assumption that all requests and responses are received by the other party reliably.

On any real network, requests and responses may be lost in transit. In order to not worry about such losses, HTTP protocol is usually layered on top of another protocol called Transmission Control Protocol (TCP). This protocol has conventions that let the other computer know whether data has been received reliably, how to detect if some data was lost and how to retransmit it, and how the receiving computer can put data back in order the sender sent it even if some data was received out of order. Note that HTTP does not have to be tied with TCP. It could run equally well on top of any other protocol that provided reliable in-order communication. Transmission Control Protocol in turn would be layered on another protocol, for example, the IP. The IP itself in turn would be layered on another set of protocols.

Most data networks are designed with several layers on top of each other, each layer consisting of one or more protocols. Classical data networking texts usually begin with an overview of a canonical protocol stack consisting of seven layers. However, the seven layers are rarely implemented in real life, so the canonical protocol stack is only of academic interest. However, terminology from the canonical protocol stack may frequently be encountered in technical networking literature. We will point out that terminology when appropriate in the rest of this chapter. The only point from the seven-layer canonical model that we will mention is that the model starts its numbering from the bottom, so that in the example of HTTP and TCP that we gave, the HTTP would correspond to a layer that is numerically higher than the layer of TCP.

It would be easier to understand modern practical computer communication protocols in terms of four layers that are shown in Figure 1.2. The bottommost layer, marked as LINK/MAC, refers to a protocol that will allow two computers to communicate with each other as long as they are physically connected to each other, for example, if there is a wire connecting them or they are within the range of wireless communication with each other. This protocol would typically correspond to layers one and two of the canonical seven-layer model. The second layer, NETWORK, refers to a protocol that will allow computers that are connected indirectly via a set of one or more

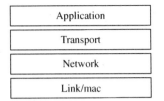

FIGURE 1.2   Data networking layers.

networks sharing a LINK/MAC layer with each other. The third layer, TRANSPORT, refers to a protocol that runs end-to-end between the two computers that are communicating across the network and allows information delivery while coping with issues such as reliability of transmission, ensuring information is delivered in sequence, and that different computers needing to communicate do not overwhelm resources in the network. The fourth layer, APPLICATION, covers any other end-to-end protocol that is required to enable communication atop transport.

At the network layer, protocols tend to fall into two major categories, circuit-switched and packet-switched. Circuit-switched protocols owe their heritage to the telephone networks, when in the old days human operators would plug-in connectors at different telephone exchanges to establish a dedicated line between two people making a phone call. In data networks using circuit-switching, a similar mechanism reserves resources for each pair of communicating computers at any intermediaries to create a dedicated communication channel. This method ensures a good quality of service but is relatively inefficient in resource usage. In packet switching, information is carried into discrete information fragments called packets, each packet carrying a header which allows intermediary nodes to send packets to the right receiver. Due to its efficiencies, most data networks today use a packet-switching paradigm even though it may be subject to fluctuations in service quality when too many packets bunch up at a single location in the network. Some network protocols use the concept of virtual circuits where a logical end-to-end circuit is created on top of a packet-switching network, which provides a tradeoff between the characteristics of both paradigms.

In order for two computers to communicate across a data network, the four layers shown in Figure 1.2 would need to be supported in a manner as shown in Figure 1.3. Assume that computer A wants to exchange data with computer B through intervening computers C and D. The information exchange requires that A and B use the same application and transport protocols. Furthermore, all the four computers need to support the same network protocol. The network protocol allows the computers A and D to exchange data with each other even though the intervening computers may be connected using a different set of LINK/MAC technologies. In the example shown in Figure 1.3,

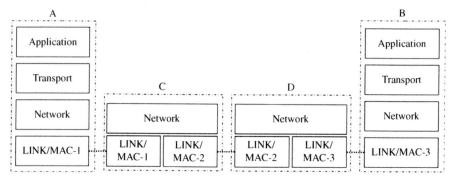

FIGURE 1.3   Data networking example.

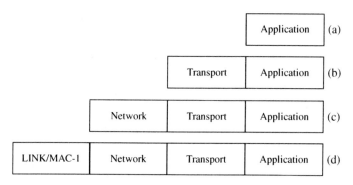

FIGURE 1.4   Data network packet structure.

each pair of computers A–C, C–D, and D–B are connected with a different type of LINK/MAC layer protocol.

When the computer A wants to send some application level data to computer B, a piece of software responsible for processing the application protocol on computer A will create that application level data. That data is marked as (a) in Figure 1.4. This data will then be passed to a piece of software on computer A responsible for transport layer processing. This software would attach some more information needed for transport protocol functions so that the data to be sent from A to B looks like the structure marked as (b) in Figure 1.4. The information needed for the transport layer is usually added at the front of the application level data and is called the transport header. The application level data is referred to as the payload. The transport header and the transport payload comprise the transport data. The transport data is then passed to a software responsible for network layer processing which attaches a network header making the data look like that of Figure 1.4(c). In this case, the transport data becomes the payload contained in the network data, which consists of a network header and the network payload (transport data).

Finally, the LINK/MAC layer-processing software will attach another header making the data look like that of Figure 1.4(d). Some portions of the preceding description may be done in hardware instead of software, but the end result is the same, producing the data format in Figure 1.4(d). This data format is what we will see if one were to take a snapshot of the data flowing from computer A to C.

Let us now examine what will happen to this packet when it reaches computer C. The LINK/MAC layer-processing software (or hardware) will strip off the LINK/MAC header recreating the data as it was in Figure 1.4(c). It will then be processed by the software responsible for the network layer at computer C, which may modify the network header. The structure would then be like that of Figure 1.4(c), but the contents of the network header could be modified and become different. When computer C sends this modified content to computer D, a new LINK/MAC header, this time corresponding to the LINK/MAC-2 protocol is attached to the data. The same sequence of steps is repeated at computer D, resulting in potential modification of network header, and the LINK/MAC header changes between every pair of computers.

The transport data is unchanged as it travels from computer A to B and processed by the transport software at computer B, which will extract the transport payload, that is, the original application data and deliver it to the software responsible for application-level processing at computer B.

The layering structure allows network communication at each layer to remain independent of the protocols used at layers above or below them. This has led to several creative combinations of network protocols to address different communication challenges that can arise in practical networks. Network protocols can be layered in many different ways, for example, an IP packet may be tunneled inside another IP packet to carry it across a network, or an IP packet may be layered on top of HTTP to cross firewall boundaries. The myriad of possible combinations of protocols has given rise to a very large and diverse set of protocol stacks.

1.3 IP NETWORKS

Although there have been many types of computer communications-networking technologies, the dominant technology in use today uses a protocol stack built upon IP. The most dominant version of this protocol that is almost universally deployed is IP Version 4, abbreviated usually as IPV4. Although there is a newer version IP Version 6 (IPV6) available in the market-place, it is yet to see significant adoption. The IPV4 protocol is the technical foundation on which the Internet and the World Wide Web (WWW) have been built. Mobile data applications are also usually based on IPV4

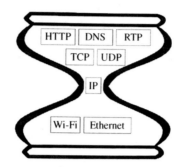

FIGURE 1.5 IP hour-glass structure.

technologies. In this book, we will simply use IP to refer to IPV4 protocol, calling out IPV6 explicitly when needed.

IP is a packet-switching network layer protocol which was developed to allow any two computers using the protocol to talk to each other, even if they happen to belong to different networks, that is, managed and administered by different organizations. Its primary purpose is to allow two networks, which may have many internal differences, to be able to exchange data effectively. In terms of the layering structure discussed in the previous section, IP would be a network layer protocol. However, as explained previously, it has been used in various other layers depending on the idiosyncrasies of specific environments.

Applications running on the IP protocol have a distribution of structure which is sometimes defined as the hourglass system, that is, if you draw a diagram of all the different types of protocols that are operational on the Internet, you will get an hourglass shape as shown in Figure 1.5. The IP protocol is at the narrow part of the hourglass. On top of it, there are a few commonly used protocols like TCP and Universal Datagram Protocol (UDP). More protocols are implemented on top of one (or both) of these protocols including the ubiquitous HTTP that defines the WWW implemented over TCP, the Domain Name Service (DNS) that is implemented on both TCP and UDP or Real-time Transfer Protocol (RTP) which is usually implemented over UDP. Above these protocols, more protocols are defined, for example, the Simple Object Access Protocol (SOAP) which defines web services is usually built atop HTTP. Application developers can create their own private protocols on top of any of the available protocols they find suitable for their purpose.

Below the IP layer are another suite of protocols which provide a way to interconnect a small set of computers. A common example is the Wi-Fi protocol, or more technically the IEEE 802.11 series of protocols. Wi-Fi is ubiquitous in homes, hotels, and airports. The Wi-Fi protocol allows a set of machines on a wireless network connected to a common access point to communicate with each other. Ethernet is another common protocol that can connect a set of machines that are connected to an Ethernet switch. Consider the situation where a Wi-Fi access point is connected to an Ethernet switch and thus acts as a bridge between a Wi-Fi network and an Ethernet network. When a computer connected to the Wi-Fi network needs to communicate with a computer that is on the Ethernet network, they both need to use a common protocol that can help them talk to each other. The IP protocol provides this function.

The bottom half of the hourglass shape indicates that a wide variety of communication protocols can be developed within smaller networks that allow a group of computers to talk to each other. The cellular communication protocols provide a similar set of protocols underneath the hourglass of IP.

The Internet consists of all the computers and other devices supporting the IP protocol that are able to talk to each other. It is a collection of several IP-based networks, operated by many different organizations, which are connected to each other at a number of exchange locations or peering points.

In an IP network, each machine has a unique address, a 32-bit identifier in IPV4, which is used to address packets to that particular machine. The different nodes in the network talk to each other to find out the right way to route packets to the correct machine. This information exchange for routing packets happens at a slower rate than packets being generated, and one of the implicit assumptions required for the correct operation of IP networks is that a machine with an IP address does not move around rapidly within the network.

In addition to the addresses, some of the machines in the IP network also have a domain name. The domain name is a hierarchical name made up with human readable characters such as http://www.chandabooks.com. If the domain name of a machine is known, then any computer can determine its IP address using a distributed system called the domain name system. The domain name provides an easier way to identify a server which needs to be accessed by many applications.

Most of the communication in an IP network falls within the paradigm of client server computing. In this paradigm, a server is a machine whose domain name or IP address is well publicized. A client is a machine which initiates communication to the server, usually by looking up the domain name of the server to get its IP address, and then sending the first packet to the server using its IP address. Once the server receives the packet, it responds back to the client and they communicate according to the protocol stack they both have in common. In the specific context of mobile data communication, normally the mobile device is the client, and the other computer it gets data from, for example a website, is the server.

Having an indirection in the domain name of the server and the IP address allows some significant flexibility in communication as well. One can develop schemes that map the same domain name to different IP addresses, for example, to a different address in each of the continents to have clients communicate to servers who are located in the same geographic neighborhood. This flexibility can be very useful to deal with the overload conditions arising due to the growth in data traffic, as explained in subsequent chapters.

One of the assumptions underlying the design of IP networks is that the location of any machine with an address within the network remains static. The manner in which packets are forwarded towards their destination reflects this design choice. There are extensions of IP communication which allow for some mobility of the machines, but despite their availability, IP remains primarily a protocol for networks where machines do not move around.

1.4 CELLULAR DATA NETWORKS

Cellular networks are also a type of computer communication network, but they evolved with the primary goal of supporting mobile users with wireless connections, very unlike IP networks which were focused on fixed computers. The cellular networks draw their name from the fact that they divide all of the area being serviced into disjoint regions called cells. Each cell is assigned a set of carrier frequencies, a portion of the electromagnetic spectrum that can be used by mobile devices in that cell to communicate with equipment in a tower servicing that cell. The frequencies can be reused in cells that are not adjacent to each other. Thus, a cellular structure allows coverage of a broad area using only a limited set of frequencies.

The tower servicing the cell can either be located in the center of the cell or on the corners of the cells. When the tower is in the center, it will use an antenna that transmits on all directions of it using the allocated frequencies. When the tower is at the edge, it will transmit on one of the allocated frequencies using a directional antenna, that is, an antenna that will transmit signals in a limited direction towards the interior of the cell. In this case, a single cell may be serviced by more than one tower.

Since users can move over from one cell to another, a system for managing hand-offs of users between cells is required in cellular networks. The specific details of how the hand-off happens depend on the communication technology which is used between a user's mobile device and the tower.

There are many protocols used within cellular networks, and some of the more common ones include General Packet Radio Service (GPRS), Universal Mobile Telecommunications System (UMTS), CDMA2000, IEEE 802.16 (Wi-Max), Long-Term Evolution (LTE), and LTE-Advanced. Each of these protocols requires a book to describe them properly. In order to not get bogged down in the specifics of these protocols, each of whom are very

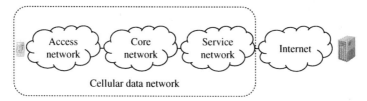

FIGURE 1.6   Components of a cellular data network.

complex, we will use a simplified description which will apply to all of these protocols in a generic manner.

The basic architecture of all types of cellular networks and the way in which they connect to the Internet can be viewed as consisting of three distinct but separate networks, an access network, a core network, and a service network. The service network will typically connect to the Internet through an IP router. Each of these components is shown in Figure 1.6.

The access network provides connectivity to the mobile devices and comprises of the equipment at the cell towers as well as other equipment required in the network for other functions, for example, controlling the operations of the radio network. Physically, the access network would consist of two distinct parts, one is the over-the-air segment which connects mobile devices to equipment at the cell tower and the other is the network which connects the cell-tower equipment to other equipment and eventually to the core network. The latter is usually referred to as the backhaul network. The backhaul network in some countries tends to be optical fiber in nature but in other countries consists of microwave.

The core network consists of devices that are responsible for authentication, access control, security, and mobility functions. As we have mentioned previously, IP networks are designed primarily to work with static machines. Within the core network, functionality is implemented so that this mobility becomes hidden from IP-based portion of the network. The exact mechanism to manage the mobility depends on the specific protocol running within the network, for example, CDMA2000 uses mobile-IP protocols to handle mobility while UMTS would tunnel packets from the mobile device to a few fixed locations in the core network called the Gateway GPRS Support Node (GGSN), in effect making the entire UMTS protocol act like a LINK/MAC protocol running underneath the IP network. The core network usually is fiber-optical in nature.[1]

---

[1] The division of cellular networks into the access network, backhaul, and core is a taxonomy we are introducing for discussion within this book. Different cellular protocols have different taxonomy which makes it challenging to discuss various cellular protocols into a single context.

The service network is an IP-based network and is the part of the Internet within control of the mobile network operator. Within the service network, the mobile network operator could run its various intermediary functions, for example, apply any transformations to convert video to fit into the form-factor required by a Smartphone, provide services such as email gateway interfaces. The set of services offered by any operator is very different and determined by the business needs of the operator.

Since many operators providing mobile services are multinational operators, the service network itself is a very complex IP network with many different regional and national networks. Even inside a single nation, the network operator may need to run a large network consisting of many different segments. The service network then interfaces with the rest of the Internet via one or more peering points.

A mobile application that is running on a mobile device would typically be oblivious to the specific type of cellular network it is running on. To the mobile application, the network is just another IP network. Thus, mobile applications are developed like any other application running on an IP network.

1.5   MOBILE APPLICATIONS

A mobile application has two components, one running on the mobile device and the other running on a server in the Internet. The data that mobile applications generate is exchanged between the mobile device and the server. In some cases, this data exchange maybe minimal. The application on the Smartphone side may run almost locally, with perhaps just an occasional check with the server side for any required updates.

A large number of mobile applications, however, rely upon the ability to exchange information with the servers. One big use of Smartphone is browsing the Internet for information. This application requires a constant exchange of data with servers on the Internet. Another popular use of Smartphone and tablets is watching video or music from sites providing such services. This requires an even larger amount of data flowing from the server on the Internet to the mobile device.

Mobile applications require data exchange to support a protocol between the Smartphone and the server for this data exchange. These protocols would sit above the IP layer in the hourglass shown in Figure 1.5, whereas the cellular protocols sit below the IP layer in the same hour-glass structure. In other words, the protocols that the mobile applications will use are transport and application-level protocols as described in Figure 1.2.

There are two commonly used transport protocols on IP networks, namely TCP and UDP. The TCP protocol gives the abstraction to two communicating machines that they have a pipe of bytes between them into which bytes can be inserted at one end and extracted from the other end in the same order without any losses. The UDP protocol gives the abstraction similar to a postal delivery services where messages with addresses can be sent by one machine and received at the other machine without any assurances about reliability or ordering. Subsequently, when security became important, a protocol called Transport Layer Security (TLS) was developed to improve the security of information being exchanged. Because TCP delivery of in-order packets can have adverse impact on the real time nature of communication, several other protocols that provide interactive voice or video delivery were developed atop UDP.

The growth of the WWW gave rise to the tremendous popularity of another protocol called HTTP. HTTP operates on top of TCP and is the protocol used by the browsers most people use. The variation of HTTP which would go on TLS instead of plain TCP is Hypertext Transfer Protocol Secure (HTTPS). Another popular feature that got added to browsers would support client-side programs written in languages such as JavaScript™. A server could send some code to the browsers which could perform some of the logic and functions right at the client side.

Given the existence and popularity of the different protocols, a mobile application developer has three possible choices when developing an application on the mobile phone:

1. Write the application using HTTP protocol for data communication, with use of JavaScript (or equivalent) for local interactions.
2. Develop a private protocol for communication needs of the application which could ride directly on either TCP or UDP, depending on the needs of the applications.
3. Use a mixed set of protocols, HTTP for some of the interactions and a private protocol for the others.

The choice made by the application developers leads to three category of mobile applications: web-based mobile applications, native mobile applications, and hybrid mobile applications. The needs of the application dictate the choice among the three modes of development.

Mobile applications built on top of IP riding on top of the cellular network are the source of mobile data exchanged among people and machines. After this brief overview of the three areas, let us examine the ecosystem of this mobile data in the next chapter.

CHAPTER 2

MOBILE DATA ECOSYSTEM

2.1 INTRODUCTION

The ecosystem of any technology consists of the set of companies, organizations, and people whose business is impacted by the technology, or whose business impacts the technology. In the ecosystem, any such entity has its own role to play, usually with a complex set of relationships to other entities in the ecosystem. When changes happen in the technology area, each of the entities in the ecosystem is impacted in some manner. Different entities in the ecosystem have different ability to react to any changes in the technology. In this chapter, we look at the mobile data ecosystem and the impact of mobile data growth on each entity within the ecosystem.

2.2 MOBILE DATA ECOSYSTEM

The mobile data ecosystem consists of many different types of companies and organizations, each performing its own specific function in collaboration or competition with other companies and organization. The different functions in the mobile data ecosystem can be divided into various roles. A role is an abstract description of an activity that is of relevance to mobile

Techniques for Surviving the Mobile Data Explosion, First Edition.
Dinesh Chandra Verma and Paridhi Verma.
© 2014 The Institute of Electrical and Electronics Engineers, Inc. Published 2014 by John Wiley & Sons, Inc.

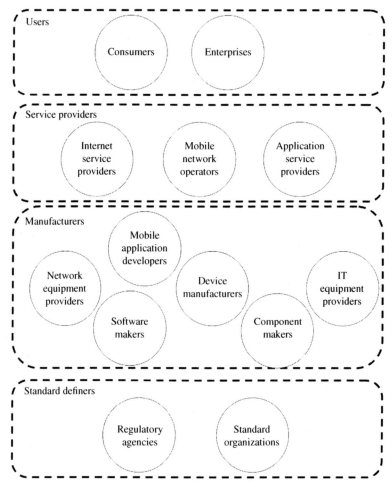

FIGURE 2.1   Mobile data ecosystem.

data-based applications. Any specific company or organization may have multiple roles in the ecosystem.

Many different roles have emerged gradually over the years in the mobile data ecosystem. Figure 2.1 provides a simplified representation of these roles. Discussing the mobile ecosystem in terms of roles instead of the individual position of different companies at the current time provides a more systematic understanding of the ecosystem. While many companies would have a single role in the ecosystem, some of them could play multiple roles in the ecosystem.

The different roles shown in Figure 2.1 are divided into four groups, shown as four different layers of roles. The actions and decisions undertaken by entities with a role in the lower layers in the figure have a significant impact on the entities in the upper layers. The reverse is also true in that the needs and requirements of the entities with a role in the upper layers are frequently taken into account and influence the decisions of the entities playing a role in the lower layers.

The bottommost layer consists of roles that have a broad effect on the manner in which other roles are performed. We have called this layer as *standard definers*. There are two main roles in this layer. The first one is that of different regulatory agencies that control different aspects of mobile data communications. These include agencies like the Federal Communications Commission (FCC) in the United States, which determine how radio frequencies are used for mobile communications, limits on electromagnetic field strength from various devices, and a variety of other regulations. Corresponding agencies with similar authority are operated by governments of other nations. Each regulatory agency makes the regulations that need to be complied with by any company, organization, or individual that comes within the jurisdiction of that regulatory agency.

The other role within the layer of standard definers is that of standard bodies. Standard bodies are associations that establish the technical specifications for running mobile networks, developing applications, or data networks. There are several such standard bodies, for example, the Internet Engineering Task Force (IETF) defines the protocols that control the interoperation of machines on the Internet, the World Wide Web Consortium (W3C) defines the standard for protocols used in web browsers and servers, the International Telecommunication Union (ITU) coordinates the use of radio spectrum across the globe and defines many standards for telephony and data communications, the 3rd Generation Partnership Project (3GPP) defines a common set of standards by which mobile phones can communicate over the air, etc. Standard bodies come in many flavors—some are associations driven by large companies, some are operated by government agencies, and some are formed by academics and technical people interested in a specific technology area. The standards that are defined and, more importantly, widely adopted by the community have strong influence on the other entities in the mobile ecosystem.

The next layer of roles consists of different types of *manufacturers*, companies that create hardware, and software that impacts mobile applications and mobile data in some way. Some of the manufacturers produce hardware components; others produce software components, while others combine these components into larger systems. Each manufacturer would sell its product to other entities in the ecosystem, including other manufacturers. From the perspective of mobile data ecosystem, we can identify three different sublayers in this layer of ecosystem. At the bottommost sublayer, we have the roles of component makers and software makers. Component makers would make hardware chips, processors, Integrated Circuits, Digital Signal Processors, and other types of components that are used in mobile devices, network equipment, and computing equipment required in the system. Software makers make software components that are used in a similar manner. The middle sublayer consists of the roles of Network Equipment Providers (NEPs), the IT equipment providers, and the mobile device makers. The topmost sublayer consists of mobile application developers.

Component makers in the mobile ecosystem consist of companies that manufacture various components that are needed to produce different types of systems. A system in this context refers to a complete box comprising of hardware components and software capabilities. Software makers in the ecosystem consist of companies that make operating systems, software libraries, and other software modules that are required by different types of systems.

Network Equipment Providers (NEPs) create boxes that allow different types of network operators to create, manage, and operate their networks. The devices offered by the NEPs implement the support for the different communication standards defined by the standard organizations.

Most of the mobile applications are written with the assumption that application will have access to the Internet. The infrastructure of the Internet requires the use of devices such as routers, network switches, and firewalls.

As described in Figure 1.1, mobile applications that send data over the network have software components running on handheld devices and the corresponding component of software running on servers. The manufacturers of handheld devices and servers have an important role in the ecosystem of mobile applications. The handheld device provides the base infrastructure that is needed to run mobile applications, and the capabilities provided by these devices have a strong influence on the amount of data a mobile application will generate. Like any other computing device, handheld devices have their basic underlying hardware, an operating system, and an assortment of applications that run on the operating system. Some of the companies provide both the hardware and operating system required for the handheld device in-house. Some of the companies produce devices with one or both of these two capabilities coming from other companies, for example, a device manufacturer may choose to offer handheld devices based on an operating system provided by another company.

Manufacturers of servers that provide computing capability for the server side of the mobile application have an important role in the mobile data ecosystem. Applications running on the server side need to accommodate access from all different types of devices, including handheld devices, desktop computers, and laptops. However, given the rapid increase in popularity of mobile applications, the workload on these servers is changing where special mechanisms may be needed to support mobile applications more effectively. Like handheld devices, server makers may opt to make servers based on their own proprietary operating system, or build servers supporting third-party operating system, for example, a version of Linux.

The number of companies that provide equipment of any type, networking gear, handheld devices, or servers, is dwarfed by the large number of companies that provide mobile applications and supporting software infrastructure required for mobile applications. Some of the supporting software infrastructures include mobile middleware, which allows applications to run across several mobile operating systems, tools for developing mobile applications more rapidly, tools that modify content to make existing applications on servers more amenable for mobile users, and infrastructure for testing mobile applications. Many of the mobile applications and enabling software components are made available to the general user population through application stores operated by some of the larger companies.

The third layer in the mobile ecosystem consists of the role of *service providers*. There are three primary roles in the layer of service providers, companies that operate mobile networks, companies that provide Internet access, and companies that provide services over the Internet. The first group of entities, the mobile network operators, consists of companies that operate cellular networks. These companies would typically buy equipment from mobile network equipment providers and offer cellular data services to their customers. Mobile network operators may also be a conduit for providing handheld devices to their customers. Other mobile network operators allow their customers to buy their own mobile devices directly from the device manufacturer.

The Internet Service Providers (ISPs) offer Internet connectivity, usually wired, to their customers. Some companies act as ISPs as well as cellular network operators, although the function would typically be performed by different units within the company. Internet connectivity generally tends to be through a wired access, for example, by means of a wire to the premises, for example, a coaxial cable or an optical fiber. However, there are a few companies that provide wireless Internet service to home or office locations as well.

The fourth and final layer in the ecosystem is that of *users*, people who use applications and services provided by the service providers. These can be divided into two groups, consumers and enterprises. Consumers are people who buy the service for individual use or for a small group of users, for example, a family. Enterprises are businesses that want to enable their employees to get access to mobile data and mobile applications. The needs and requirements of an enterprise may differ substantially from that of a consumer. In general, an enterprise may need more security and control of the applications that run on the Smartphones of their employees, may want to standardize on the set of applications that run on the phone, and may want to track and manage the way their employees use their handheld devices. These considerations are generally not present in the case of consumers.

Within the ecosystem, an entity may have more than one role. The manufacturer of a server may also have the role of an enterprise. A standard setting organization may have multiple premises and use mobile data services like an enterprise. An application service provider might also be the manufacturer of an operating system. Sometimes, different roles are played by different divisions within a company, and sometimes the difference may not be very sharply defined.

As mobile data and mobile applications keep growing, each of the different players in the ecosystem would have a different set of roles and responsibilities. Each player would be impacted in a different manner, and the approaches they would use to address this growth in mobile data would also be different.

2.3 MOBILE DATA GROWTH

Since much of this book deals with the issue of mobile data growth, it is worth looking at the nature of mobile data growth and the drivers behind the growth of mobile data. Several companies [1,2] study the characteristics of users of mobile as well as fixed access networks, and by combining the results of those studies, we can get an understanding of the nature of mobile data and its growth.

To a large extent, the growth in mobile data has been caused by the uptake of Smartphones and tablets and by users switching to these devices instead of using their laptops or personal computers. In 2011, sites that were originally intended for users accessing the Internet via personal computers had about two thirds of their users accessing them from mobile devices [3]. Social networking sites have more than a third of their users accessing them from mobile devices, and the trend is similar for almost all other sites.

Another major change resulting in an increase of user data has been the shifts in type of applications popular among different users. Several studies [4,5] track usage patterns among Internet users regularly. According to the data collected by several such studies, the predominant application on the Internet around 2000 was web browsing. On the other hand, the predominant application on the Internet in 2011 was real-time entertainment such as watching movies streamed from online services. The amount of data downloaded by a user when watching a half-hour of a streaming video is significantly larger than that of the same user surfing the web for the same amount of time. The growth in popularity of audio- and video-intensive applications is one of the major drivers of the usage of mobile data.

The nature of mobile data tends to show strong variations among different geographies, since the nature of the consumers and users there tend to be very different. In the United States and Western Europe, Smartphones, tablets, and specialized electronic book readers are very popular, replacing a large fraction of basic phone users, and are the primary source for mobile data. Some of the reports published in the industry [6] indicate that an operator providing access to a new Smartphone can experience more than 14 times the data volume growth before such an access was granted, with the obvious conclusion that Smartphone users on the average generate more than 14 times the data volume of a basic phone. In the U.S. market by end of 2009, Smartphones accounted for about half of all mobile data [3]. On the other hand, in the rest of the world, Smartphones tend to be a much smaller fraction of overall mobile users. In these geographies, laptops accessing mobile broadband data services tend to be the primary drivers of mobile data. usage behavior in these geographies shows a different mix of traffic usage, with peer-to-peer file sharing accounting for a larger share of bandwidth usage.

As per currently published reports [6], the top applications driving mobile bandwidth consumption in the United States are downloads of audio and video, web browsing, and access to social networking sites. Within Western Europe, web browsing edges out real-time entertainment. In Asia Pacific and Africa, peer-to-peer file sharing applications are one of the largest applications contributing to the network traffic.

The nature of applications that drive mobile data growth has a strong implication on the strategies that can be used for addressing the pressures created by this growth.

2.4 WHERE IS THE BOTTLENECK?

As mobile data grows, the question arises as to where is the bottleneck caused by the growth of the data in mobile networks. The answer tends to differ across different geographies, but some common trends and patterns can be observed.

Mobile data flowing between the mobile device and the server traverses three types of intervening networks, the cellular network, the Internet, and the data center network. The data center network usually has sufficient capacity with high-speed switching gear and is not a bottleneck. The Internet is a large swath of individual networks that interconnect at various locations. While not all the links of the Internet are necessarily high-speed, most of it consists of wired optical fiber where more bandwidth can be added as needed. Some of the segments of the Internet, for example, the transcontinental links, may have lower bandwidth than other segments but are generally provisioned adequately. The cellular network, in itself, however has a different story.

As shown in Figure 1.6, the cellular network itself consists of three segments, the access network, the core network, and the service network. The service network itself is like any other section of the Internet, comprised of a wired infrastructure, and not a significant bottleneck. The core network is in a similar situation.

The access network, however, is a very different story. It consists of two main portions, the radio network that connects the cell-tower equipment to the mobile devices and the backhaul that connects the cell-tower equipment to the core network infrastructure. Both of these portions of the network may become overloaded due to the increase in mobile data traffic.

Within the radio network, the bandwidth that is available is a function of the spectrum available for communication between the mobile devices and the cell-tower equipment. The portion of the spectrum that can be used for mobile data communication is limited and usually available through expensive and limited licenses from the government in each country. The spectrum available for mobile data communication is a fixed resource, with the only option to have additional spectrum is to have it be released for such use from the appropriate government regulatory agency. With the anticipated growth in mobile data traffic, this bottleneck is not likely to be eliminated anytime soon.

The other part of the access network, the backhaul, also suffers from a congestion problem, although that is much more a question of economics than actual resource shortage. The backhaul connects cell-tower equipment to equipment in the core network. There are three predominant technologies used to provide backhaul access, microwave, copper, and fiber optics. Microwave links use a licensed portion of the spectrum to communicate wirelessly. These types of links are the predominant mode of backhaul in Western Europe. In contrast to wireless communication provided by microwave, copper and optical fiber provides wired communication technologies for backhaul. Copper technology uses traditional coaxial copper cables to provide connectivity, while fiber-optics is an alternative that provides higher speeds at lower costs. Microwave installation and operation is more commonly found in areas where providing wired access, which requires putting down cables, rights of access, or digging up the earth to put in underground cables, is not feasible or too expensive. However, microwaves can only operate at limited bandwidths, typically under 100 Mbps, but newer microwave technology can go up to 350 Mbps and even 1 Gbps. In contrast, achieving higher rates using either copper or optical fibers is routine with fiber having a much higher available rate than copper.

Due to historical reasons, copper-based backhaul is the predominant mode in North America to cell phone access towers. Replacing them with optical fibers is an expensive proposition for cellular operators. However, with the increasing bandwidth demands due to customers, such a replacement may have to happen gradually. Upgrading capacities in areas where fiber-optic access exists is much easier. New emerging markets such as China, where cell phone operators rolled out the network in recent times, have rolled out fiber optics to the base stations. Backhaul access is not a bottleneck in such areas.

The core network of the cellular network operator and the Internet are largely based on wired communication links based on fiber-optic technology. In these cases, there are no fundamental technology limits in the amount of bandwidth that can be provisioned. The delays across the Internet tend to be fairly low, but there are still additional points of concern created due to the variation of traffic load on the network. Some of the causes of delays in the Internet include congestion at peering points where different ISPs come together to interconnect their networks, inefficiencies introduced due to routing protocol dynamics, and momentary disruptions in network connectivity [7]. In some portions of the Internet, delays may arise due to inefficiencies of access links like Digital Subscriber Line (DSL) [8]. Nevertheless, the impact of such delays would be less pronounced overall than the delay in the access segment of the cellular network.

2.5   IMPACT OF MOBILE DATA GROWTH ON THE ECOSYSTEM

The growth of mobile data has different implications on the different players in the ecosystem. In this section, we look at the impact on each of the players in the mobile ecosystem due to the growth of mobile data.

The user community in the mobile ecosystem is one of the drivers of the growth of mobile data. We have identified two types of users in the ecosystem, consumers and enterprises. Consumers typically are single users or a small group (like a family) while enterprises consist of larger sets of users (few tens to several thousands). Generally speaking, the growth and availability of mobile devices is beneficial for both consumers and enterprise users since it allows them connectivity to Internet-based resources from any place. Enterprise users can also access the resources available to them only by a secure virtual private network to their enterprise intranet.

The convenience of using the mobile phone to do various tasks from anywhere and anytime is tremendous. Whether users want to entertain themselves by watching videos and surfing the net or conduct some work on the Internet, mobile phones improve the ease of use significantly. The two issues that adversely impact this convenience are security and quality of experience.

Security risks increase with mobile usage because mobile devices can be more easily lost than other types of computing devices. This can lead to loss of privacy and identity theft for consumers, resulting in significant inconvenience and financial losses. The risks to enterprise users are even higher. Enterprise users may lose valuable data stored on mobile devices, which may cause loss of intellectual property, and depending on the type of data that is lost, expose the enterprise to significant financial and operational risk. Thus, as mobile devices become more widespread, enterprises have to deal with issues that arise when mobile devices get lost, and the resulting exposure.

A congested network can cause serious degradation in the quality of experience obtained by the user. The convenience and ease of mobile applications can easily change to frustration if the network has large delays. This can have a significant impact on the usage of mobile data and devices. With bandwidth being limited, consumers may see an increase in their prices for data usage. A consumer worried about the data communications bill may not use the mobile applications and devices as much as they would like to. Similarly, enterprise users need to figure out how to manage access from their mobile device to the enterprise services when the network link connecting the users to the servers in the enterprise is congested.

If we look at the other constituency in the ecosystem, those of the service providers, we can see the obvious impact of a congested network on both mobile network operators and application service providers. Mobile network operators are running a congested network, thereby incurring decreased satisfaction of their customers. Upgrading the network infrastructure to cope with the increased demand will require significant capital investment. Passing portions of these upgrade costs to consumers is likely to adversely impact customer satisfaction. Electing not to invest in the upgrade is not an option either, since the mobile network operator who offers the best connectivity is going to get the most users. Mobile network operators have the pressing task of delivering the best customer user experience at the lowest possible cost.

Application service providers create applications running on servers in the Internet and provide access to them to Internet users. They are faced with a similar challenge of managing their user experience when users are communicating over a congested network. Unlike mobile network operators, application service providers do not have the possibility of upgrading and changing congestion on the underlying network. Nevertheless, they need to determine how to provide any application in a manner that can cope best with a congested network.

Among the set of various service providers, ISPs may appear to be impacted the least by the growth of mobile data. Mobile data eventually becomes data on the Internet, and the technologies to support data on the wired Internet are sufficiently advanced so as not to pose a problem in terms of capacity or congestion. Nevertheless, mobile users have a behavior that is remarkably different from those of desktop users. Thus, as more users become mobile, the nature of traffic on the Internet may shift and change, which will have an impact on the way ISPs run and operate their networks.

The techniques that can be used to implement the strategies for coping with mobile data growth by the different service providers have to be implemented by equipment manufacturers. Another set of techniques has to be implemented by developers of the mobile applications and supported within the operating systems and platforms that are used to run and transport mobile applications. These include approaches such as deploying new networking technologies, creating new communication protocols, designing applications differently, and creating new services in the network. As new technologies emerge and are supported by the different players in the manufacturing space, the relative fortunes of an individual entity may change depending on their decision to adopt and support the new technologies.

Some of these techniques will span multiple manufacturers and organizations and would need to be put into standards. This will impact the standards development. Government regulatory agencies have a special role in tackling this challenge. They are the ones who control the scarce spectrum resource that is available on the air, and their decision to release some or part of the spectrum currently unused or used for other purposes can have a significant impact on the severity of network congestion caused by the growth of mobile applications.

Most of the techniques that can be used to deal with the growth of mobile traffic are described in subsequent chapters of this book.

CHAPTER 3

AN OVERVIEW OF TECHNIQUES FOR BANDWIDTH OPTIMIZATION

3.1 INTRODUCTION

As mentioned in Chapter 2, mobile data traverses through a variety of different networks as information is exchanged between a mobile device and a server located in a data center connected to the Internet. Each of these networks are supported and managed by network operators. Each network operator tries to maintain the maximum quality of experience of users accessing different data services on their network, while managing costs and improving its own profitability. There are three main challenges that any network operator faces as traffic increases: (i) how to get the maximum bandwidth out of its existing network that may be getting congested, (ii) how to reduce its cost of operations as the data growth happens, and (iii) how to make more money and profit from the data flowing on the network.

This chapter looks at the first challenge—the task of getting more data through a network with a limited capacity—and discusses the various techniques that can be used to address the challenge. The techniques are presented in the context of a simplified model of a computer communication network. They can then be mapped onto the special cases for applying within different segments of the networks that may be congested for mobile data. In the first section of this chapter, we introduce that idealized model of the

Techniques for Surviving the Mobile Data Explosion, First Edition.
Dinesh Chandra Verma and Paridhi Verma.
© 2014 The Institute of Electrical and Electronics Engineers, Inc. Published 2014 by John Wiley & Sons, Inc.

network and discuss the effect of network congestion on user quality of experience. The subsequent sections discuss various approaches that can be used to squeeze more data through a limited capacity network conforming to the idealized model.

3.2 NETWORK MODEL

Although network protocols and network structures can be very complex, the principles governing their performance and capacity can be explained using a simplified network model. The simplified model is unlikely to be found in a practical network, but it allows an effective benchmark to compare the relative merits and benefits of the different techniques for bandwidth optimization. Figure 3.1 shows the simplified model of the network that we will use to explain the various bandwidth optimization techniques.

As shown in the figure, the model of the network consists of a sender and a receiver. Data packets flow from the sender to the receiver. The sender and receiver are connected through a network, which consists of several intermediary nodes, with communication links connecting the sender, the receiver, and the various intermediary nodes. One of the links in between the sender and the receiver is the bottleneck link, that is, it is the link that has the least possible capacity among all of the links. This link is the shaded one shown in the figure.

In order to understand the concept of a congested link, let us look at the relationship between the delay encountered by packets flowing between the sender and the receiver with respect to the capacity of the bottleneck link. Let us assume that the bottleneck link is capable of transferring packets at the rate of C bits per second and that the sender needs to send packets to the receiver at the rate of R bits per second. In order for the delay between the sender and receiver to be finite, it is required that R be less than C. If we define U as the ratio R/C, then U must be less than 1.

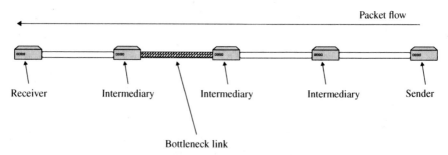

FIGURE 3.1   The simplified network model.

In the aforementioned network, there will be some amount of delay between the time the sender sends a packet in the network and the time the receiver gets that packet. This delay has a significant impact on the quality of experience of a user getting any content from the server. As an example, using most of the common networking protocols, the time to download a file from the server would be impacted significantly by this delay. If we were to plot that delay against U, we will usually get a curve like that shown in Figure 3.2. The delay experienced depends on how many bits are pending to be transmitted on the bottleneck link. This amount is negligible if the rate at which bits can be transmitted on the bottleneck link is much higher than the rate at which bits arrive at the bottleneck link. However, when bits arrive at a rate higher than the rate at which they could be processed, the number of bits pending transmission adds up and can result in large delays. This behavior is reflected in the plot, which shows that the delay becomes very large as U becomes closer to 1. There is a value of U beyond which the delay increases rapidly as U increases and under which the delay tends to be relatively small. This marks the knee of the delay curve.

Problems arise in the network when the total volume of traffic that is required to be transported in the network is such that the value of U will go into the area beyond the knee. In order to avoid getting into this problematic situation, one has to either reduce the amount of the traffic R that is being generated or increase the amount of capacity C that is available. The increases in capacity would generally require approaches specific to the type of network that is deployed.

In this chapter, we look at approaches that will reduce the total amount of traffic R that is generated into the network. In all of these approaches, the network needs to support traffic of R being sent from sender to receiver.

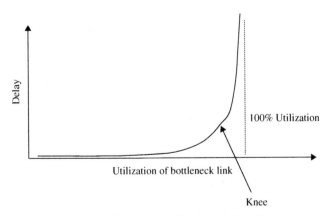

FIGURE 3.2 Delay-utilization relationship.

It does so by implementing techniques that reduce the traffic from R to an amount R' such that the ratio R'/U falls below the knee of the delay curve.

3.3 OBJECT CACHING

Object caching reduces the amount of traffic going over the bottleneck link by placing some of the content being accessed over the link closer to the receiver. In the idealized diagram that is shown in Figure 3.1, the goal is to change the location of the sender to avoid the bottleneck link. Instead of receiving data from the sender, the receiver gets it from a proxy for the sender. The proxy is located at a point in the network so that the traffic from the receiver to the proxy does not traverse the bottleneck link.

The idealized network shown in Figure 3.1 changes to the network depicted in Figure 3.3, when object caching is used. A proxy for the sender is located between the sender and the receiver, and positioned so that packets from the proxy avoid the bottleneck when being sent to the receiver. If there are multiple bottleneck links in the network, more than one proxy will be needed.

If some of the packets that were flowing from the server instead came from the proxy as shown in Figure 3.3, the load on the bottleneck link can be reduced. The effectiveness of the scheme depends on the relative magnitude of flows that can happen from the proxy instead of the original sender. Suppose a fraction f of the total bytes that need to be sent to the receiver can come from the proxy, then the utilization of the bottleneck link changes from U to fU. Since f is less than 1, the utilization will shift away from the knee of the curve, as shown in Figure 3.2.

How are such proxies created within the network? In most common types of network, flows usually happen in a transactional manner following a client–server paradigm. The server waits for requests from the client. The client locates a server within the network and then makes a request to get some data object to the server. The server starts sending the requested data

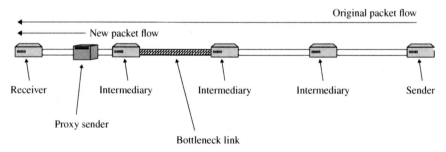

FIGURE 3.3 Object caching in the simplified network model.

object to the client. Data is communicated between the client and the server using protocols they both understand.

Proxies can be introduced in the path of the request either (i) when the client is trying to locate a server or (ii) when the client is making a request to a located server. When the client is looking for the server, they can be provided with the name of the proxy server instead of the real server. Alternatively, the network can identify requests being made to a server, and instead direct those requests to the proxy server.

When requests are directed to the proxy, only a subset of the requests can be satisfied by the proxy. In a caching proxy, the proxy would forward the first request for any data object to the original server, acting like a client to the original server. When the response to the request is received, the proxy keeps a copy of the data object locally. If subsequently a request is made for the same object, the proxy can respond using the local copy of the object.

Sometimes, a data object can be updated with a newer version at the server while the proxy has an older version. In this case, the response provided by the proxy to a client could be outdated. To avoid this situation, one of two schemes can be used. In the first scheme, the proxy checks with the server on every request if the data object has been modified. This check, which involves only comparing the time-stamps of when an object has been changed, takes considerably less bandwidth than retrieving the actual object. The other scheme requires the server to send a notification to the proxy when the data object changes. The original server needs to keep track of various proxies who have a copy of the data object and notify them of the updated version.

Object caching has been used widely in computer networks for many different types of protocols. Perhaps the most common case of object caching is in the context of the Hypertext Transfer Protocol (HTTP) protocol, using caching proxies for static pages of the Internet. A comprehensive survey of many types of web-caching proxies and the techniques deployed within them can be found in Reference 9. In general, the cache is transparent to both the client and the server, a goal that can be attained if the protocol between the client and the proxy is the same as that between the proxy and the server. Support for caching is built into the HTTP protocol [10] and allows the caching proxy to check for updated content, as well as provide hints from the server to the proxy whether a specific content can be cached or not. As an example, dynamically generated pages can be hinted as being uncacheable by the server. A variety of techniques that allow servers to invalidate stale content in the proxy caches have also been developed.

In a network, there may be multiple caching proxies instead of a single one. When these proxies are distributed in the network, and an intelligent scheme to route the requests to the proxy that can best service that request from the client is deployed, the proxy caches together constitute a content distribution network [11]. Many commonly deployed content distribution networks in the Internet use the Domain Name Service mechanism to direct requests to the most suitable proxy on a client request.

3.4   OBJECT COMPRESSION

In any type of data communication, the goal is to transfer some data object between a sender and a receiver. One way to reduce the amount of bytes required to be transferred over the bottleneck link is to compress the object before transmission. This approach requires that the protocol between both the sender and the receiver be able to identify and support compressed objects.

There are a variety of algorithms and approaches that are available to compress a data object and reduce the amount of bytes that are required to transmit it. Compression algorithms are classified into the two broad categories of lossless compression and lossy compression. In lossless compression, data is compressed so that the entire original data object can be reconstructed without any errors. In lossy compression, portions of the original data object are considered less important, and a better reduction is obtained by using approaches where those portions are discarded. Lossless compression is used for general data objects, while lossy compression is used for objects where some amount of loss can be tolerated, for example, audio or video content which humans are able to understand even if the content is slightly degraded.

The most popular lossless compression algorithms belong to the Lempel-Ziv family [12,13]. These algorithms look for repeated patterns or sequences occurring in the content and replace them with a smaller representation of the same on the second and subsequent occurrences. As the algorithm processes a data object, it will create a table that would store the bit-patterns to the corresponding shorter representations. The table is generated dynamically, depending on the contents of the object being compressed. There are several other lossless compression algorithms [14,15] that have been presented in the literature.

Lossy compression is usually more efficient than lossless compression. Many of the common techniques for representing images, audio and video files, for example, JPEG, MPEG, and MP3, deploy variations of lossy compression techniques. In lossy compression, the idea is to drop portions of the original data that are considered unimportant. There are various approaches and algorithms that can achieve this goal. See References 14 and 15 for a survey of both lossy and lossless data compression algorithms.

The effectiveness of the compression scheme on the utilization of the network depends on the reduction in size that can be obtained on the contents being transferred. If the compression algorithms reduces on the average an object of size S into a compressed object of size cS, then the utilization would change to cU from the original utilization of U. This will result in the effective utilization of the bottleneck shifting to the left on the curve shown in Figure 3.2.

3.5   PACKET COMPRESSION

Although data compression is useful in reducing the number of bytes being transferred on a bottleneck link, it suffers from one practical problem. In order for it to work well, both the sender and receiver have to be aware of the fact that compressed content is being transferred. For a large number of protocols that may have already been developed assuming the transfer of uncompressed content, this scheme requires some drastic modifications. That modification on the different servers and receivers may not be practical from a cost perspective in many cases. Packet compression allows a large portion of the benefits of compression to be obtained without requiring changes to application level software.

The use of a packet compressor in the network shown in Figure 3.1 would result in a network that looks like the one shown in Figure 3.2. There are two boxes inserted on either side of the bottleneck link. These two boxes need not be immediately next to the bottleneck link, but one is needed on both sides of the bottleneck link. A packet being sent by the sender is changed to a smaller packet by one of the boxes when traversing the bottleneck link, and restored to the original packet by the other box. The box reducing the size of the packet is the packet compressor shown in the figure while the box restoring the size of the packet back is the packet decompressor (Figure 3.4).

Packet compression techniques are variants on the concept of general data compression. In most of the common communication protocols, data is

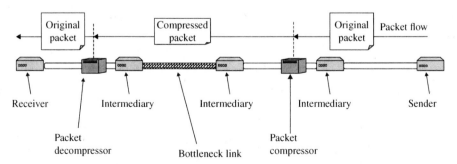

FIGURE 3.4   Packet compression in the simplified network model.

transmitted by dividing it into smaller units called packets. The part of the data carried in each packet is the payload of the packet. In addition to the payload, each packet contains a header, which is useful for routing the packets and reassembling the payload at the receiver side. When compressing the contents of the packet, either or both of the payload and the header can be compressed to save on bytes transferred.

Payload compression uses the same algorithms as lossless data compression for bulk data. However, since the compression is being done online between two different devices, some modifications to the general compression techniques are needed. It is generally done using a dictionary-based compression algorithm. A dictionary, which contains the mapping of some common types of occurring bit patterns to a more efficiently coded pattern, is used at both the sender and the receiver. The dictionary is used to replace the frequently occurring contents of the packets with a smaller code. Depending on the frequency of occurrence of the patterns in the dictionary, such compression schemes can be very effective in reducing the content of information.

A relatively new variation on packet compression is the technique of byte-caching. In byte-caching [16], the dictionary is synchronized between two devices, and the frequently occurring content is replaced with a small cryptographic hash. Instead of having a predefined dictionary, byte-cache builds the dictionary dynamically based on the contents of the packets that are flowing between two pairs of devices. The bytes that are to be matched with specific hash-codes are maintained in a cache at each of the devices, hence the name byte-cache. Byte-cache can thus achieve bandwidth savings that packet by packet compression cannot. If the same content is being transmitted to multiple receivers, byte-caching can identify that duplicate content is being sent across multiple streams and replace a large number of such packet content with much shorter hashes.

Other variations on dictionary-based payload compression include maintaining multiple dictionaries at each point and selecting them on a packet by packet basis based on the content, or creating estimates for predicting future byte-streams based on current content [17,18], which may allow for more efficiency.

Header compression relies on the fact that the standard headers in Transmission Control Protocol (TCP) and Internet Protocol (IP), the protocols used for communication in much of the present day communication networks, impose a significant bit of overhead in many situations. Under conditions of normal use, the TCP and IP headers is at least 40 bytes long. When used for communication on bandwidth constrained links, this may impose a significant overhead, especially for packets with small payloads. Header compression techniques have been standardized via various Internet Request for Comments (RFCs) [19–21], which describe different techniques for reducing the header size. The general idea behind these techniques is that many of the fields in the packet header do not change frequently on some links. As an example, if a source and destination are communicating on a serial link, these two fields are going to remain the same regardless of the packet exchange. In these cases, one can use a streamlined header that does not require specifying these fields. Similarly, if some information is changing slowly, a different scheme may be used to communicate it rather than transmitting it in its entirety. These techniques result in new types of headers for the IP packets. A performance evaluation and high-level survey of techniques for packet payload and header compression can be found in Reference 22.

3.6   FLOW SHARING

Flow sharing can be viewed as a variation of packet compression that employs the knowledge of the content that is being sent on a congested link. As mentioned previously, an object cache can be used to check if the same content is being accessed by two different users, and subsequent requests can be sent a cached copy of the response to the first request. However, it requires that the content be static. It does not work if multiple people are requesting information to content that is dynamically generated and being sent out.

There are some common examples where this situation can arise. Sports events are broadcast live on the Internet and many people see them as they are broadcast. Both audio and video are commonly broadcast in this manner and are formats that require a higher bandwidth than normal text. Other examples where similar situations arise is interactive video meetings with multiple participants, where typically the same live content is being transferred to multiple recipients. Popular lectures being streamed over the Internet, or corporate executive broadcasts, are also examples of this situation. Although these applications constitute a small fraction of all applications, the large presence of video content in them makes them a big user of network bandwidth.

Object caching will not work in this case since the content is being dynamically generated. Packet compression may not be as effective since content that goes to different receivers does not usually get packetized in the same way. Variations of the packet compression, for example, byte caching that can work across packet boundaries to identify content, may be effective at reducing the bandwidth required by these types of applications, but they require significant computation and caching across multiple packet boundaries to make the determination of shared content. Flow sharing can obtain the same goal at a higher level and provides a scheme to bridge the benefits of packet compression with the efficiency of object caching.

A flow in the network is a stream of packets flowing between two devices. The basic idea in flow sharing is to identify two or more flows that are carrying the same content and transmit the content only on one of those. The basic concept in flow sharing is illustrated in Figure 3.5 and Figure 3.6, respectively. Figure 3.5 shows the original configuration with a single sender talking to multiple receivers. There are many copies of the same content flowing over the bottleneck link over the different flows.

Figure 3.6 shows the same set of senders and receivers when flow sharing is used within the system. In this situation, two boxes are located on each side of the bottleneck link, much like the situation in the case of packet compression. The box located on the sender end takes the flows that have the same content and puts them into a single flow. The corresponding box on the receiver side takes the single flow and divides them into multiple flows, one per receiver.

In IP-based networks, flows can typically be identified using five fields in the header of each IP packets, its source and destination address, the upper level protocol whose data it is carrying, and the source and destination ports the upper layer protocol is using. These five fields are also called the

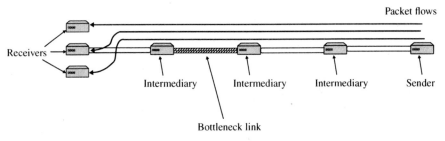

FIGURE 3.5   Multiple flows in the simplified network model.

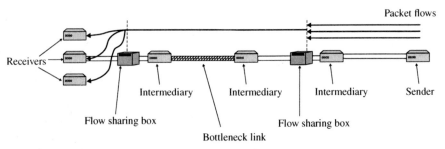

FIGURE 3.6   Flow sharing in the simplified network model.

five-tuple. If two flows, identified by their five-tuples are carrying the same content between two points, they can be replaced by a single flow.

The use of flow sharing requires the use of two devices, like that in the packet compression schemes. At one of this pair of devices, a set of rules is used to identify which of the multiple flows are carrying the same content. When two or more such flows are identified, the device at the other end is signaled to indicate the similarity of these flows. When both the devices agree to compress the flows based on the signaling mechanism, the first device can start suppressing all but one of those network flows. The second device would then recreate multiple flows from the content of a single flow.

Flow sharing can be performed at different layers of the network protocol. At the IP protocol layer, one way to share flows is by leveraging IP multicast [23]. In IP multicast, routers automatically provide a single flow provided the application has been written to be aware of multicast. IP multicast is not always supported in the networks, but an intervening network link with congestion can still reap the benefits of flow sharing by using a scheme such as Automatic Multicast without Tunnels, or AMT [24]. AMT allows multicast flows to be identified and transported as such on a unicast network. Such flow sharing can also be implemented at application layers by leveraging application layer proxies [25]. These proxies allow multicast at an application layer by intercepting and sharing network flows as needed.

The use of application-level proxies for sharing flows has another advantage. Instead of having two boxes for flow sharing, application-level proxies can interpret the communication protocol used between clients and servers. They can manipulate their behavior so that only one box is required for the purpose of sharing flows. The setup of such a flow-sharing device would be as shown in Figure 3.7. The flow-sharing box shown in the figure can interpret application level requests and responses. As a result, it can determine which requests are ones that can be satisfied locally from an existing flow in the system and which require going back to the server. As an example, consider an application on a mobile device used to watch the live streaming broadcast of a game. In addition to the broadcast video

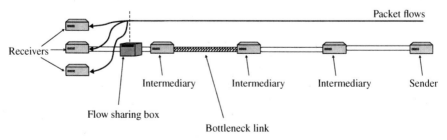

FIGURE 3.7 Flowing sharing with single box approach.

stream, the application may also have a second panel, which could be showing advertisements that are personalized to better appeal to the person watching the program. An application level proxy would understand which requests from the application pertain to the broadcast video stream, which can be satisfied locally from the flow manager, and which ones are for the personalized advertisement panel, which need to go to the server across the bottleneck link.

The savings in bandwidth and reduction of the bottleneck link using flow sharing can be tremendous when used for bandwidth-heavy applications like streaming video, live broadcasts, and web conferences. The bandwidth savings on the bottleneck link becomes more significant as the number of receivers sharing the same flow increases.

3.7   CONTENT TRANSFORMATION

Some objects being transferred on the mobile network can be made available in different formats, with each format using a different amount of bandwidth on the network. Video streams, for example, can be encoded at different rates [26]. Many common types of video encoding schemes [26,27] produce video that contains a base layer and several enhancement layers. The video quality is acceptable when only the base layer is sent to the receiver and becomes better as more and more enhancement layers are used. This allows a sender to control the data that a video stream consumes, depending on the rate that is available.

Similar types of variable rate encoding can also be done for other types of content intended for human consumption, for example, images. An image could be reduced in size by representing it using a lower resolution, for example, reduce the pixels per inch used to represent the image, or by scaling down its size while keeping the same resolution. Depending on the amount of bandwidth that is available, different versions of the picture can be transmitted.

A proxy in the network or the sender of information can use this capability to decide the version of object that ought to be transferred on the network. Figure 3.8 shows how content transformation reduces the amount of bandwidth on the simplified network model used in this chapter. The figure assumes that the transformation is being done by a proxy. The proxy intercepts the high-bandwidth content being sent by the sender and converts it to content that will use lower bandwidth.

When this type of adaptation is done depending on the characteristics of the network, we get an adaptive transmission scheme for the network. An adaptive scheme tries to get to the suitable type of content into the network depending on the amount of bandwidth available. Several adaptive techniques that can be used at the application level are known [28] and used in

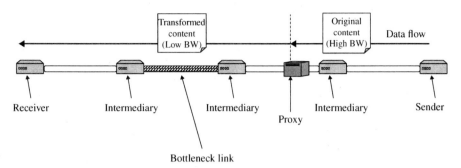

FIGURE 3.8   Content transformation in the simplified network model.

many different implementations. A judicious selection of the right type of content to transfer can result in significant saving of network bandwidth without serious degradation in user quality of experience.

3.8 JUST-IN-TIME TRANSMISSION

Any piece of information that is copied over a bottleneck can be used in one of two modes, which we can roughly call the download mode and the streaming mode. In the download mode, the entire piece of information needs to be available at the receiver before it can be used. An example of the download mode would be a PowerPoint presentation that is downloaded from the Internet. Until the entire presentation is loaded, this file cannot be viewed by the receiver. The streaming mode is for information and can be used even if part of it is available at the receiver while the rest is being downloaded. Video played from Internet sites are an example of streaming mode. Once a few seconds worth of video is available at the receiver, it can be played while the rest of the video is still in transit.

Applications that use the streaming mode generally maintain a lead time at the receiver side. This lead time is the amount of time the receiver can keep on using the data that is already received by it. In most of the video players used on the Internet, the lead time is usually indicated by means of a status bar in the client. When the network is fast, the lead time can be quite substantial. In many cases, an entire video clip can be downloaded completely even when the receiver has only been able to watch a small fraction of it.

Many people do not watch a streaming movie completely but stop it after watching the first few seconds. In those cases, if a significant portion of the movie was not watched, the unwatched bits at the receiver are discarded. From the goal of maximizing useful data transfer on a congested link, a large lead time is counterproductive. If a user switches the video off after a few short seconds, the amount of data that is already loaded at the receiver needs to be discarded.

The goal of just-in-time transmission is to put an upper limit on how much lead time needs to be maintained at the receiver. The lead time needs to be chosen carefully. It should be large enough to prevent stalling when playing out, but short enough so that a prematurely terminated video does not result in a large number of unused bytes being discarded. The techniques used to ensure that the lead time is within some bounded zone are collectively known as just-in-time transmission.

Just-in-time transmission to manipulate the amount of data that is pending at any receiver requires the receiver to report back on the lead time it has maintained to either the sender or to a proxy in the network. The proxy or the sender, located on the other side of the bottleneck link, can adjust the transmission so that the appropriate lead time is not exceeded.

In order to estimate the effectiveness of just-in-time transmission, let us consider the case where a large fraction of bandwidth on the bottleneck link is due to multiple users watching streaming videos. Assume that the average video is about 10 minutes in length, but the average user only watches 2 minutes of a video on the average. By keeping the lead time to be no more than 1 minute, the average user will download 3 minutes of the video instead of the complete 10 minutes, thus reducing the bandwidth demand by 70%.

3.9 RATE CONTROL

When something is in short supply, for example, gasoline or food supplies, it is not unusual for governments to mandate rations or quotas on its usage. The intention is to prevent a few people from hogging resources and maintain a more or less equitable share of the scarce resource.

Rate control, in a similar way, is the process of limiting the amount of bandwidth any user can get within the network. Suppose we have 100 users trying to access a link that has a capacity of 100 Mbps. On the average, each of them should be getting a fair share of 1 Mbps. However, because of the various idiosyncrasies of different network protocols, some of them may be getting more share of the bandwidth than others. Some users, by making different kinds of tweaks on their system, would be able to get more than their fair share at the expense of others.

Let us examine a common mechanism that is used (or rather abused) by some of the products available in the industry. The TCP protocol is designed so that different TCP-based communication flows are very polite to each other. If there are n TCP connections flowing through a congested link, like in the scenario shown in Figure 3.1, each of them will adjust their sending rates till each of the connections will get $1/n^{th}$ of the bandwidth of the congested link. Normally, if there are 100 applications running across the link, each would open one TCP connection for its data transfer, and each will get an equal share of the congested network link. However, if a developer wants to get an unfair advantage for his set of applications, so that they appear to run faster, he could open two TCP connections instead of just using a single connection. Now, there are 101 connections sharing the link, but his application has 2/101 share of the congested link, double that of everyone else. He need not open only one additional connection; he can open 10 or 20 connections to get even more of the share of the bottleneck link. By being greedy in a world of nice applications, a developer can get their applications to run much faster.

If every application developer started being greedy and using the same trick, nobody would have an advantage. The situation will be much worse for everyone. Everybody will lose because they have to maintain more connections in their application and will have more complexity. For this reason, very few application developers resort to such behavior. However, there are always a few people who may be quite willing to use such tricks.

When some applications are greedy and the others are nice, the greedy applications will crowd over the nice applications. There are many ways to make an application greedy, and in a large distributed network, there is a fair chance that some users or applications will revert to schemes that will get them a larger share of a constrained resource.

Rate control is the scheme by which different packet flows going into a link are limited to use no more than a specific amount of the bandwidth on the link. The rate control mechanisms usually involve a packet scheduling mechanism at each of the link in the network. The scheduling mechanisms will impose an upper limit on the amount of resource that any individual network packet flow can take.

3.10 SERVICE DIFFERENTIATION

When one has more users than can be handled in the available amount of bandwidth, one can divide them into multiple groups of different priority. One can then give preferential treatment to one group over the other. The idea is to preserve the service to the group that has the higher priority at the expense of service degradation to the group that has a lower priority. This type of prioritization of one set of users over another is called service differentiation.

There are many different ways in which such a difference in service quality can be administered. Instead of giving everyone a share of the reduced bandwidth equally when there is an overload situation, one group may get a higher share of the bandwidth than the other. The difference in the bandwidth available allows better service quality to the users who have the higher rate allocation.

Some networks, especially those that are connection oriented, have a signaling mechanism that is required in order to set up a connection or admit a new user into the network. In these types of networks, service differentiation can be offered at the time of signaling. The process is called admission control, and some users are admitted in preference over others. When there is an overload situation, the requests to join the network are refused when they come from the users belonging to the lower priority group. This allows the users in the higher priority group to join the network with a higher probability of success. In some cases, users with a lower priority that may have already joined the group may be pre-empted when another user belonging to a higher priority group makes an attempt to join the network.

A similar differentiation in priority can be offered in networks at a time of packet transfer instead at the time of signaling. The packets belonging to higher priority users are given preferential treatment compared to the packets belonging to the lower priority users.

Both rate control and service differentiation are closely tied to the concept of policy. Policy determines which application, user, or network flow gets mapped to a level of service or defines specific limits on the network bandwidth that can be used. Different types of networking gear have their mechanisms for enforcing policies. The more important aspect of policy control, however, is an administrative decision as to which class of users is more important than others, which is generally defined on nontechnical criteria.

CHAPTER 4

AN OVERVIEW OF TECHNIQUES FOR COST REDUCTION

As mentioned in the previous chapter, network operators want to (i) get the maximum bandwidth out of their existing network which may be getting congested, (ii) reduce their cost of operations as the data growth happens, and (iii) make profit from the data flowing on the network. In this chapter, we look at some of the approaches that help in reducing the cost of operation for the mobile network operator.

In the preceding chapter, we looked at the aspect of pushing more data through a bottleneck link without the need to upgrade it with additional capacity. These approaches are one way to reduce the cost of operation, since they defer the expenses associated with getting a faster network bandwidth. Nevertheless, these are not the only approaches that reduce the cost of operation. In this chapter, we will look at some of the other approaches that can help in reducing the costs of running a mobile data network.

4.1 INTRODUCTION

The growth of mobile data has put additional expenses on every constituency in the mobile ecosystem that we described in Chapter 1. Consumers are paying more for the new generation of Smartphones and their data plans.

Techniques for Surviving the Mobile Data Explosion, First Edition.
Dinesh Chandra Verma and Paridhi Verma.
© 2014 The Institute of Electrical and Electronics Engineers, Inc. Published 2014 by John Wiley & Sons, Inc.

Enterprises and application service providers need to put in additional infrastructure (servers and software) to allow their employees to access information and enterprise services over mobile devices. And mobile network operators need to put in additional investment to upgrade their networks. Even the bandwidth minimization techniques that we discussed in the previous chapter require insertion of new devices, which come at a cost.

Mobile network operators find themselves in a somewhat precarious situation. They need to invest in the upgrade of their network to support the higher data rates required for mobile devices, which is a significant expense due to the scale of their network. If their networks are not fast enough, they risk losing customers to other competing mobile network operators. At the same time, most of the profitable services enabled by the growth of mobile devices are provided by application service providers who are running their services on the Internet. For most of these services, mobile network operators do not have any significant advantage over application service providers in offering the same service. Thus, it is imperative for mobile network operators to look for ways to reduce their expenses.

In order to understand how a network operator can reduce its costs, it is helpful to explore the structure of the network as the technology for networking improves and new applications are developed. New technologies for improving network bandwidth do come up all the time. At the same time, new bandwidth-hungry applications that can benefit from the availability of higher bandwidth are created in response to the growth of higher bandwidth. This, in turn, can cause congestion in portions of the network.

As a result of the continuous cycles of technologies offering more bandwidth, and the growth of resulting bandwidth-hungry applications, most networks have a structure that is depicted in Figure 4.1. The figure shows several clusters that are well-connected internally, but the communications link connecting the different clusters tend to be overloaded.

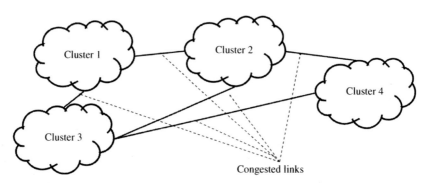

FIGURE 4.1   Clustered structure of networks.

The entire network has pockets of good connectivity with a few points where the network links do not have sufficient capacity.

The reasons for the existence of bottleneck links are usually associated with cost, geography, and administrative boundaries. In multinational networks, the networks within the same country may be very fast, but communication capacity outside the national boundaries may be limited due to the costs associated with them. Frequently, within the Internet infrastructure, the links providing peering between different Internet Service Providers have less capacity than the networks that they operate internally within themselves. In other cases, costs, technical limitations, and business agreements cause the creation of links that may not have sufficient capacity. In many European cellular carriers, the mobile cell towers in nonurban areas are connected via microwave links. While one can get adequate capacity by upgrading the connection to that of an optical fiber, the costs associated with running fiber make that option unviable. In some cases, the network lines connecting some clusters may be leased or acquired from another operator. In those cases, the cost of leasing the bandwidth may be the factor preventing the operator from acquiring more bandwidth.

Thus, due to a variety of reasons, the network infrastructure acquires the structure shown in Figure 4.1—clusters of well-connected networks that have some bottlenecks at specific parts. Some approaches for cost reduction will work well for the portion of networks within a well-connected cluster, while other approaches will be more suitable for the portions that have a bandwidth bottleneck.

4.2 INFRASTRUCTURE SHARING

The infrastructure required to operate a mobile data network includes a part of the wireless spectrum, the cell towers and equipment installed there, the network links connecting the cell towers to other locations in the network, and the networking gear required at various locations within the different networks. This infrastructure is expensive, and is a significant component for the net expense required to provide mobile data services. A similar infrastructure is required for other types of networks.

One way to reduce the cost of operation is to share the infrastructure among more than one network operator. In that way, the cost is divided among different operators, and each has to pay only a fraction of the total cost. As an example, two or more companies may decide to share the infrastructure of a cell tower, and both can place their equipment in that cell tower.

The sharing of infrastructure can range from a simple sharing of facilities, for example, two operators sharing space in a cell tower to more comprehensive sharing of different links and network equipment. Beyond just sharing space in a facility, two operators may decide to share a part of their network infrastructure. On the wireless side, this may involve the operators sharing the same network equipment that may operate in two spectrum bands, one band for each of the operators. Beyond this level of sharing, the two operators may also opt to share the same spectrum band with subscribers of both operators accessing the same spectrum. On the wired side, two operators may share the same backhaul network that connects their cell-tower equipment to the core of the network, or may share portions of the wired network as well.

There is a geographic dimension to sharing of infrastructure. One operator may have a significant infrastructure in one region of the world, while another operator may have a significant infrastructure in another region. These two operators can share facilities with each other, with one of them leveraging the facilities in one region, and the relationship getting reversed in another region. In a third region, both of these operators may be sharing the infrastructure provided by a third operator. In some cases, an operator may have a presence in the majority of an area and may sublease its infra-structure to other operators for a rental fee.

As two operators share more of their infrastructure, they need to give up some of the control they have over the shared infrastructure. As an example, one operator may be worried about relatively lax security standards of the other and concerned that security lapses of the other operator may increase its liability. As the level of sharing increases, the amount of control of an individual operator decreases.

4.3 VIRTUALIZATION

A technology that alleviates some of the concerns related to sharing is that of virtualization. Virtualization is the technique to make something look like an independent copy of itself. Let us say we have a physical widget, where the widget could be a computer, a memory stick, or a network link. A virtual widget behaves in the same manner as the physical widget, but it would be created by using only a portion of the physical widget, by combining several physical widgets, or by using a slightly different type of physical widget.

As an example, a network link of 100 Mbps capacity can be made to look like 10 virtual links of 10 Mbps each so that the traffic on each of the virtual links is separated. Similarly, 10 physical links of 100 Mbps capacity can be made to look like a virtual link of 1 Gbps capacity. A physical network link that supports a given network protocol, for example, Ethernet, can be augmented with software drivers and/or physical connectors at its edges to become a virtual link that supports another protocol, for example, Fiber Channel. Similarly, the concepts of virtual machines in servers that subdivide a server, or virtual memory that creates a memory of a larger size than the physical memory, are commonly used in the industry.

When the network infrastructure is shared, virtualization technologies can be used to significantly alleviate the concerns associated with sharing. A virtual network can be separated from another virtual network. Virtualization technologies that provide sufficient isolation allow each operator to manage its virtual component while enjoying lowered costs due to sharing of the infrastructure. The concept of virtualization can be applied to radio access networks, backhaul networks, core networks, and computers that provide the functions required in the network.

Virtualization enables an operator to provide part of its infrastructure to other operators for sharing in a secure manner. The exact form in which the sharing may happen may be different, depending on the geographical presence and business relationships among different operators.

4.4 CONSOLIDATION

The principle of consolidation to reduce costs is a variation on the general concept of "economies of scale." In general, as one uses bigger systems to offer any service, the cost per unit decreases. A large farm has lower costs per unit of crop produced compared to a small farm. A large factory can produce widgets at a lower cost per widget compared to a smaller factory. Even though the larger factory or large farm costs much more to operate, both in terms of fixed cost as well as operating costs, the costs are lower per unit when they are amortized over a larger number of units.

The situation with computers and communications is no different. Larger computers have a lower cost per instruction than smaller computers even though they are more expensive in absolute terms. Fatter network pipes have a lower cost per unit of bandwidth than thinner network pipes. In order to reduce the cost of operation, consolidating operations whenever possible on larger servers and fatter bandwidth pipes will reduce the overall cost of operation.

Such consolidation is very much possible in well-connected network clusters in the structure of the network as shown in Figure 4.1. Each of these clusters is made up of different types of devices, which perform various types of functions. Suppose the function in the cluster is performed by devices connected in a topology as shown in Figure 4.2. This is a hierarchical topology of devices that provides connectivity between other clusters with whom the network capacity may be constrained. The other clusters are shown as clouds

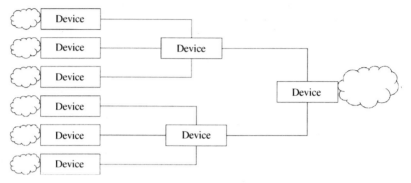

FIGURE 4.2   Hierarchical structure in a well-connected cluster.

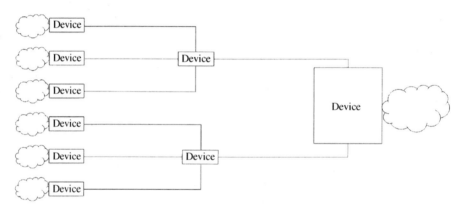

FIGURE 4.3   Alternate structure in a well-connected cluster.

in Figure 4.2. For the sake of illustration, we assume that the devices at each location are identical in size and capacity.

Figure 4.3 shows an alternate configuration with the same topology, but with the difference that all of the devices but one are replaced with miniature or smaller versions of the same. The smaller versions will not perform all of the functions that the normal-sized version of the devices shown in Figure 4.2. Their main function is to make sure that the packets coming to them are forwarded to a larger and more powerful version of the device. This larger device is shown at the root of the tree in the hierarchical structure shown in Figure 4.3. However, in principle, the larger device could have been located at any place within the network as long as the smaller devices had the capability to route packets to and from the larger device.

If the smaller devices were cheaper than the normal devices, then the total cost of the system shown in Figure 4.3 would be significantly less than the total cost of the system in Figure 4.2. Suppose there are N devices within

Upper most layer

Bottom most layer

Device

FIGURE 4.4 Structure of a device.

the network cluster, with C_s, C_n, and C_l denoting the cost of the small, normal, and large devices. The cost here could be the capital expense associated with the devices, the operational expense associated with the devices, or any combination of the two. The cost of the configuration shown in Figure 4.2 will be NC_n while the cost of the configuration shown in Figure 4.3 will be $(N-1)\,C_s + C_l$. The configuration shown in Figure 4.3 will be cheaper if $(N-1)$ $(C_n - C_s) > C_l - C_s$. In other words, the additional costs of getting one single large device are offset by the savings attained in replacing the others by smaller devices. In most large-scale networking environments, the number of devices N is very large, and the savings obtained by consolidation could be substantial.

In order to obtain this cost savings, consolidation must be possible without sacrificing the quality of network performance. A key requirement for this consolidation to happen is to have high-speed connectivity among the cluster. As mentioned in Chapter 1, networking functions are invariably structured in a layered manner. Thus, the functions contained in each of the devices can be viewed as being those required to support the different layers of the networking stack. The structure of a normal device as shown in Figure 4.2 is shown in Figure 4.4. The bottommost layer, shown with a dashed box in the figure, is normally the one used to interconnect all of the devices in the cluster into a network. The upper layers may consist of several different functions, depending on the complexity of their protocols.

When device consolidation is done, the functions at the bottommost layer of the communication protocol are all that have an absolute need to remain in the small devices. As an example, the cellular protocols for the wired portion of networks are usually designed to run atop ATM or IP protocols. If there are sufficiently high-speed links available, the processing of the cellular protocols can be concentrated into a large device, while the majority of the cluster consists of only IP routers or ATM switches.

Consolidation to reduce cost has been used in several projects and is applicable to different parts of communications and computing infrastructure. Consolidation can be applied to radio access networks [28] as well as core network [29] components of cellular networks.

4.5 IT USAGE IN NETWORKS

Telecommunication operators typically divide their infrastructure into two distinct areas, the network and the Information Technology (IT) infrastructure. The network is responsible for carrying voice and data and runs on special purpose network equipment. The IT infrastructure supports billing and record keeping and consists of software systems running on computer servers. The computer servers that are used in many industries besides the telecommunication networks tend to be much cheaper from a processing capacity per dollar perspective. One way to reduce costs for network operation would be to run networking operations as software on IT servers. The question is whether one can make this transformation happen without losing the reliability and performance offered by specialized networking hardware.

As mentioned in Chapter 1, most of the networking infrastructure is defined in layers, which build upon one another in functionality. There are four common layers of networking capability that were described in Figure 1.2. Of these layers, the bottom two layers are present in all of the networking gear, whereas the top layers are typically used among the end-points of communications. The upper two layers are typically run as software running on IT devices, for example, PCs, laptops, and servers, while the bottom two would typically run in special boxes. The IP layer would be processed within the software on these special boxes, whereas the MAC layer would be processed in specialized hardware within the special boxes.

In practice, it is not uncommon to find many more than four layers of protocol in operation. The layered architecture lends itself to recursive stacking, and the application layer of one type of network technology is layered on top of another type of network technology. As a result, if one were to capture a packet in the middle of a network and examine it, it would not be unusual to find seven or eight layers of nested protocol headers in the packet.

In general, it is easier to convert the functionality implemented in upper layer of the protocol stack than the lower layers into software. The lower layers of the protocols, specially the physical layer, are harder to migrate to software because they have more stringent timing requirements and require specialized digital signal processing operations. However, as processors increase in capacity, it becomes possible to implement more and more of the lower layers as software running on standard IT systems.

The exact division boundary in the protocol stack between the hardware and software depends upon the cost of implementing a function in software versus hardware. If there is a high volume of demand for components that are required to process a layer of protocol, and that function is stable, for example, implementing a widely adopted standard, the volumes of production are likely to make that function cheaper to do in hardware. On the other hand, if a protocol function is subject to rapid change, requires more flexibility, and has low volume demands, that function is preferably done in software. As processing capacity increases over time, it becomes viable to push more functions in software.

Several research projects have investigated the option to run the bottommost protocol, the radio protocol in software [30,31]. These projects have shown that it is technically feasible to run all of the network protocols on standard IT servers. However, the lowest cost solution may not necessarily be a software implementation of all the protocols. In many cases, the lowest cost solution may be obtained by only putting some of the upper layer protocols in software, leaving the others to hardware cards or integrated hardware subsystems.

Within the cellular network industry, it is not uncommon to find special boxes that implement all the protocol layers, including the application layers. These special boxes tend to be more expensive than standard IT servers. Cost benefits can often be gained by migrating from special hardware to software running on standard IT systems. In such a system, most of the upper layers of the protocol processing will be done in software, while the bottom layers which may not be cost-efficient to perform in software will be done by means of special hardware cards attached to IT servers, or by means of a hybrid systems architecture that uses a IT processor in conjunction with specialized hardware for some of the lower layer networking functions.

In some networks, protocol functions are divided into control operations and data operations. Control operations are functions required for initialization and operation for the network, for example, exchange of messages to determine how information can be routed within the network. Data operations are functions required for actual transfer of information. Control operations, in general, would use less computational and communication resources to perform than data operations. As a result, control operations in many types of networks can be implemented as software on IT servers, instead of being implemented as special hardware in a distributed networked environment. Such an implementation can reduce costs and complexity of network equipment.

SECTION 2

TECHNIQUES FOR MOBILE NETWORK OPERATORS

CHAPTER 5

BANDWIDTH OPTIMIZATION AND COST REDUCTION IN THE RADIO ACCESS NETWORK

In this chapter, we will look at the specific challenges that a mobile network operator has to face in the area of radio access network. While we have looked at some general approaches in the previous two chapters, this chapter applies those general approaches in the specific context of a radio access network.

5.1 INTRODUCTION

As mentioned in Chapter 2, the primary bottleneck due to the mobile data growth lies in the radio access network. The approaches to alleviate that bandwidth bottleneck are described in this chapter.

At a high level, the techniques described in this chapter can be divided into four broad categories. The first category consists of approaches that can obtain additional bandwidth for the users on the radio access network by adding, modifying, or upgrading the radio access network. The second category consists of approaches that leverage bandwidth available from any additional network connected to the radio access network. The third category consists of approaches that try to regulate, control, and manage the usage

Techniques for Surviving the Mobile Data Explosion, First Edition.
Dinesh Chandra Verma and Paridhi Verma.
© 2014 The Institute of Electrical and Electronics Engineers, Inc. Published 2014 by John Wiley & Sons, Inc.

of bandwidth on the radio access network. The final category consists of techniques that try to suppress the demand of bandwidth from the users.

5.2 UPGRADING THE RAN

The most obvious approach to alleviate the shortage of bandwidth on the radio access network is to get more bandwidth. The simplest way to do this would be to get more electromagnetic spectrum for use on the radio access network. The data throughput that can be achieved on the air is directly proportional to the amount of electromagnetic spectrum available for communication. Adding extra spectrum can ease the bandwidth crunch of the radio access network significantly.

The spectrum available for mobile data communication in any geography is controlled by a government agency with appropriate jurisdiction. In order to release more spectrum for mobile data communications, the spectrum needs to be available. Given the physical limitation of available spectrum, it may mean taking it away from some other alternative usage, such as transmission of television signals or military communications. Even if the government agency manages to release some additional spectrum, it is not likely to be very frequent or sufficient to satisfy the growing demand for mobile data. The costs of acquiring the spectrum are also significant, licensing costs running into billions of dollars in most countries.

Even if additional spectrum is made available, it is a physically limited quantity and will run out eventually. In that case, the mobile network operator has to find a way to obtain maximum data bandwidth from the spectrum available to it. In most cases, it would lead to changes in the design of its radio access network. Some of the possible design options that can improve the available data throughput are described in the following subsections.

5.2.1 Technology Upgrade

One of the ways to get more bandwidth out of the same spectrum is to use a different technology for communication, one which has a higher spectral efficiency. Spectral efficiency is a measure of how many bits per second of data can be carried in a given frequency range of the electromagnetic spectrum. As a general rule, the spectral efficiency of newer protocols tends to be better than the spectral efficiency of the older protocols. This means that switching over to a newer technology can get more data throughput on the same spectrum.

There are many different technologies that can be used on cellular networks to carry data. The number of protocols operating in the wireless space shows the same chaotic number of diverse network protocols that existed in the wired communication networks from 1970 to 1990. During that period, a number of protocols such as Appletalk, Systems Network Architecture (SNA), Advance Peer-to-Peer Networking (APPN), Internetwork Packet Exchange (IPX), and Internet Protocol (IP) were used and supported by different companies and in general did not interoperate with each other. Since the mid 1990s, the IP protocol overtook all of the other protocols in popularity and became the dominant protocol, reducing the others to rather isolated islands of usage. That type of convergence to a single protocol has still not happened in cellular data networks. As a result, there are a large number of protocols used in cellular networks, each of which would take several books to describe completely. In the next few paragraphs, we would present a simplified way to understand these protocols and their relation to other protocols.

There are three prominent families of cellular data network protocols that are used in various areas around the world today. These three families are the 3GPP family, CDMA family, and the IEEE family. Each family of cellular data protocols have a set of protocols that are roughly classified into generations, 2G, 3G, and 4G, where the G stands for generation and the number indicates an increasing level of sophistication in the radio network protocol technology. With increasing generation number, the technology is able to support a higher data throughput than the previous protocols in the same family.

Figure 5.1 shows the major cellular data protocols in the different families. Each family represents a standard organization or association that defines cellular protocols. The 3rd Generation Partnership Project (3GPP) is one such association that has produced a family of protocols. Another association called the 3rd Generation Partnership Project 2 (3GPP2), which was independent and competing with 3GPP association, produced a different set of standards, which is shown as the code division multiple access (CDMA) family in Figure 5.1. We could have called it the 3GPP2 family, but the CDMA family avoids the obvious naming confusion. The third major set

	3GPP	CDMA	IEEE
2G	GSM, EDGE, GPRS	CDMA2000	
3G	WCDMA, UMTS	EV-DO Rel. 0	
3G+	HSPA, HSPA+, LTE	EV-DO Rel. A EV-DO Rel. B	WiMax
4G	LTE-advanced		WiMax-advanced

FIGURE 5.1   Cellular protocol families.

of protocols for cellular data networks has been produced by the Institute of Electrical and Electronic Engineers (IEEE).

In the figure, the 3G protocols are broken into a further subset of 3.5G, which contains protocols that are faster than the 3G ones but not quite as fast as the corresponding 4G protocols in the same family. The protocols from different families in the same generation have sustainable data throughputs that are roughly comparable.

The 3GPP protocols are the ones most commonly deployed in the world, with the Universal Mobile Telecommunications System (UMTS) or Wideband Code Division Multiple Access (W-CDMA) networks making up the bulk of the cellular networks in different countries. W-CDMA networks come in a few flavors depending on the exact nature of the modulation techniques used on the air. The next generation of such networks includes the High-Speed Packet Access (HSPA) and HSPA+ protocol suite and the advanced version called Long-Term Evolution (LTE). A higher capacity protocol called LTE-Advanced has also been specified by the 3GPP association. Currently, many mobile network operators use W-CDMA protocol and are in the process of migrating to LTE.

The 3GPP2 association brought out a set of competing standards that are known as the CDMA2000 protocol in the 2G version. Subsequently, for data communication, the Enhanced Voice-Data Only (Ev-DO) standards were developed, which have been produced as three releases, Release 0, Release A, and Release B, each providing a higher possible throughput.

The IEEE has produced the Worldwide Interoperability for Microwave Access (WiMax) specifications, which were deployed by some selected carries in the United States and some other countries. The advanced version of WiMax supports higher data rates comparable to that of the LTE-Advanced.

The reasons for the existence of these many standard organizations are based on politics and history, a subject too complex to deal within the scope of this book. It is worth mentioning that there are a few cellular protocols that lie outside these three main families. We are ignoring these protocols in the context of this book since they do not add much value to the main thesis of this book. We are mentioning them only to clarify the point that the protocols enumerated in Figure 5.1 do not cover the entire universe of different cellular protocols.

Upgrading the radio access network to a higher generation protocol in the same family will get more bandwidth to the users. However, there are significant costs associated with such an upgrade, including the costs of changing the systems at the cell towers, replacing the handsets of the users to ones that support the new protocol, and upgrading any components of the system that may get bottlenecked with the upgrades. As a result, technology upgrades in most mobile data networks happen slowly over a prolonged period of time.

5.2.2 High-Density RANs

Suppose a mobile network operator is unable or unwilling to change the technology of its radio access network, perhaps because replacing all the handsets of the users is a time-consuming process. It needs to look for approaches that can be used to get more bandwidth while using the same technology on the air. High-Density Radio Access Networks provide one way to achieve this goal.

As mentioned in Chapter 1, the radio access networks are designed so that their access area is divided into regular areas called cells, and the cells are serviced by one or more cell towers. The amount of spectrum that is available within the cell is able to sustain a given amount of data throughput, which is dependent on the nature of the protocol used within the cell.

Suppose we reduce the size of a cell so that each side of the cell is half that of the original. The area of the new cell will be one fourth that of the original cell. If the original cell covered an area where a couple of hundred users were active, then the new cell only needs to support a fourth of those users, or about fifty users. The same available bandwidth is now shared among a smaller set of users, and each user can get a larger fraction of the capacity that is available in the cell.

Dividing the coverage area of the radio access network into smaller sized cells is in general referred to as a high-density RAN deployment. The original cell is referred to as a macrocell or a large cell, while the smaller cells are called variously as microcells or picocells. Small cells or high-density RAN deployments result in more bandwidth by dividing the available bandwidth among a smaller number of users.

Figure 5.2 shows how the high-density RAN deployment system works. Suppose you need to cover a rectangular area in a city to provide coverage. One way to do it will be to have one large cell that covers the entire rectangular area. Mobile device coverage planning is usually shown as hexagonal cells, primarily because hexagonal coverage areas can be placed next to each other

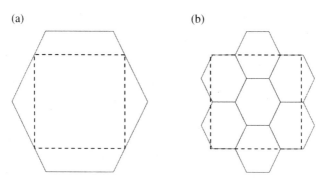

FIGURE 5.2 High-density RAN: (a) macrocell and (b) microcell.

to cover an entire plane. The diagram in Figure 5.2a shows how a large cell can be used to cover the entire area. The diagram in Figure 5.2b shows how one can cover the same area using cells that are a third in size. The same area is now covered using seven such cells.

There is another benefit of using smaller cells. The power required by the equipment at cell tower increases as the distance covered by it increases. The power required for transmission is proportional to the square of the distance that needs to be covered. As a result, smaller cells tend to consume much less power than larger cells. The equipment required at the smaller cell size therefore tends to be smaller and less power hungry than that at a macrocell tower. Looking again at Figure 5.2, the power required at each of the smaller microcells would be about 1/9th of the power required at the macrocell that is required to cover the area shown in the figure. Since there are only seven microcells required to cover the area, the net transmission power consumption required of the system would typically decrease.

Despite the reduction in the size and cost of the equipment required at a cell tower in a high-density deployment, the total cost of all the small cells can easily add up. In order to reduce the cost of high-density deployments, a version of the consolidation approach described in Chapter 4 can be used. The base-station equipment can be divided into two portions, a Radio Unit (RU) and a Base Band Unit (BBU). The RU is normally responsible for on the air transmission and reception function while the BBUt performs the other functions required for protocol processing. The traditional base-station equipment function is provided by one RU and one BBU. The radio units can be taken out of the base-station unit and converted into a remote radio unit (RRU). The RRU is a very small and compact piece of equipment, which does not need a lot of space or much infrastructure and is installed at the small cell sites. Many of the RRUs are connected to a central BBU, which provides the role of acting as a BBU to all of these RRUs. The total cost of the system is much lower than installing a complete RRU–BBU unit at each of the cell sites.

Figure 5.3 shows how the distributed equipment infrastructure works. Figure 5.3a shows the normal equipment that will be required at each of the cell-tower sites. The equipment needs to have the functions that implement both the capabilities of the RU (or RRU) and the BBU. In real equipments, the function may not be as neatly separated out as shown in the diagrams, but both those capabilities need to be present. When the two functions are broken out, one of the cell sites takes over the task of processing the BBU functions for all of the other cell sites. Figure 5.3b shows that by means of a larger BBU at the central site. The other cell sites only have the RRU which will be smaller and cheaper and require less infrastructure support.

An extension of this model for distributed base-station architecture has been proposed in the experimental cloud-RAN architecture [29,30].

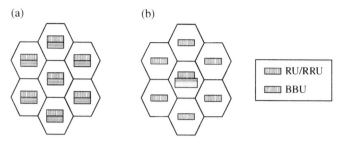

FIGURE 5.3 Distributed equipment infrastructure: (a) normal and (b) distributed.

In this architecture, the RRU at the cell towers are connected via high-speed optical fibers to a central location, which consists of normal computers typically used in data centers. Most of the functions that are implemented using specialized electronics are implemented as software running on computers, with even radio functions implemented in software.

5.2.3 Multihop Cellular Networks

The high-density deployment provides a way to get additional bandwidth by introducing additional static number of small cell sites. However, there is no need to restrict the additional cell sites to be laid out in a fixed pattern or in a place where there is wired access. One way to get additional bandwidth in the Radio Access Network is to have base-station equipment that can be moved as needed to areas where additional bandwidth is needed. These base stations can communicate with each other or to a fixed base station using wireless interconnectivity themselves. This leads to the concept of what is known as a multihop cellular network [32].

In traditional cellular networks, cell towers are sometimes mounted onto a truck or trailer. This leads to a cell-site on wheels, often referred to as a COW. A COW can be rolled out to provide additional bandwidth when there is a need to provide additional cellular communication capacity to a group of users that may form for a limited period of time, for example, when a popular sporting event where many people are likely to congregate is being held. A COW can be used to augment capacity in this manner for a brief period.

When a COW is used, the communication between a mobile device and the base station at the cell tower goes through two wireless hops, one from the mobile to the COW and the other from the COW to the cell tower. This concept can be generalized to an architecture in which the mobile may reach the cell tower through not one but multiple hops. The difference between a normal cellular network and the multihop cellular network is shown in Figure 5.4.

Figure 5.4a shows how many devices moving around in an area covered by a single cell tower would communicate with the equipment at the cell

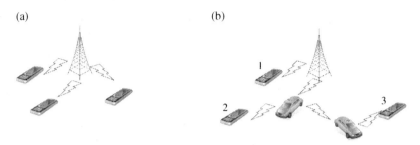

FIGURE 5.4   Multihop cellular network: (a) single-hop and (b) multihop.

tower. Each of the different mobile devices connects directly to the cell tower equipment. Figure 5.4b shows how the mobile devices would connect to the cell tower in a multihop network. In this case, a mobile may go to the cell tower indirectly using intermediary equipment that may be mounted in a taxi, a bus, or other type of fixed or mobile device. One of the mobiles (mobile 1) connects directly to the cell tower, while another goes through two hops (mobile 2) and yet another goes through three hops (mobile 3).

Multihop cellular networks can be viewed as several COWs that are interconnected together in a dynamic wireless mesh. They are also known as hybrid networks. A mobile may reach a base station that is connected to a wired infrastructure by means of multiple hops, with its packets passing through one or more COWs along the way. The position of different COWs may be relatively fixed with respect to each other, or variable. In the latter case, the different COWs are moving relative to each other, with more complexity in managing the connectivity topology and routing of information among them.

While traditional COWs tend to be relatively large, the concept of multihop cellular network also envisions the equipment mounted on taxis or buses that are operating in a metropolitan area. The mobile nodes provide an alternative method to increase coverage for people who are in the neighborhood. This approach can work well for urban areas where the dense population and the high number of taxis and buses make the concept practical.

The concept of having multiple hops at the edge of the cellular network is attractive and can be used to augment existing cellular capacity. Although it has been studied in various academic publications, it has not yet been deployed widely.

5.2.4 Network in a Box

A different concept for providing dynamic capacity in wireless networks is provided by various start-ups whose offerings can best be described as a mobile cellular network in a box. As mentioned in Chapter 1, a wireless network consists of two parts, the radio access network and the core network. The various functions of the two portions of the networks are provided by various devices installed in the network. As an example, in the Universal Mobile Telecommunications System (UMTS) network, four types of devices are used to provide the total functions of the mobile wireless networks, these devices being called nodeB, Radio Network Controllers (RNC), Serving GPRS Support Node (SGSN), and Gateway GPRS Service Node (GGSN), where GPRS is an acronym for General Packet Radio Service. Other radio network standards have similar appliances but they are called by different names.

Several companies implement the functionality provided by these devices as software modules and allow all of the devices required for implementing the complete wireless network as a software system. The radio protocols that go on air are usually supported by means of a radio card or another appliance that can be installed on a traditional computer system. The radio card, coupled with the software running on computers, allows one to convert any sufficiently powerful computer into a complete network, offering a network in a box. This provisioning of network in a box using software capabilities is an instance of IT usage in networks mentioned in Chapter 4.

Such complete networks can be used to augment the capacity of existing networks at a lower cost than obtaining various types of devices. They can also be used for creating additional capacity as needed in a very small factor for dedicated applications on a wireless network, such as implementing a solution for monitoring electric grids over wireless networks.

5.3   LEVERAGING ADDITIONAL BANDWIDTH

The solutions in the previous section looked at obtaining more bandwidth by making changes to the structure of the radio access network or by introducing additional devices within the radio access network. In many cases, there may be additional networks available in the vicinity of users and can be leveraged to obtain additional bandwidth. In this section, we look at the opportunistic use of this network capacity. Techniques that fall into this category include femtocells and Wi-Fi offload in the access network.

5.3.1   Femtocells

A femtocell [33] can be viewed as an example of a small cell site, but with an important difference. While the other cell sites, a macrocell, microcell, or picocell, are connected to a wired infrastructure owned by the operator of the

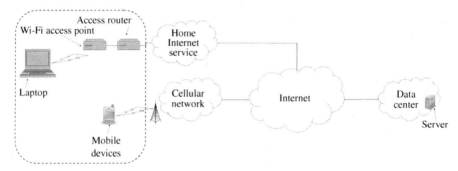

FIGURE 5.5   Typical home access.

radio access network, the femtocell connects to the Internet service which is subscribed to by the users of cellular service.

A large number of people today have wired Internet connectivity at their homes. This connectivity is obtained through a subscription to an Internet Service Provider. The typical setup of Internet access at a home or small office is shown in the dotted oval on the left side of Figure 5.5. The right-hand side shows the network connectivity outside the home. A mobile device like a cell phone at the home (or outside the home) would connect to the cellular network via the equipment at the cell tower. Most homes today also have computers at home that can access servers on the Internet. The typical configuration has Internet access provided by means of an access router. The access router would connect to the portion of the Internet that is operated by the company providing home Internet service. The access router would connect to the provider network via the infrastructure supported by the provider, Integrated Services Digital Network (ISDN), Asymmetric Digital Subscriber Line (ADSL), coaxial cable, or fiber optics. It is quite common to have the access router be connected initially to a Wi-Fi access point. Many devices in the home, for example, computers, laptops, set-top boxes, or game consoles connect to the Internet using Wi-Fi as the first hop of the connection (Figure 5.6).

The paths used by the mobile device and a laptop at home would typically be different and are shown in Figure 5.5. When the mobile device would need to access a server at a data center connected to the Internet, it would transmit the data over the radio access network to the cell tower, which then gets relayed via the cellular network to the Internet and onward to the data center. That path is shown as path A in the figure. A laptop accessing the same server would use the path shown as path B in the figure. If the radio access network is congested, then the mobile device might as well use path B, which is available at hand. This is where femtocells come in.

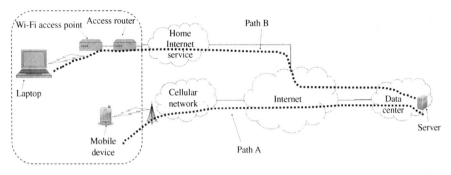

FIGURE 5.6   Communication paths in typical home access.

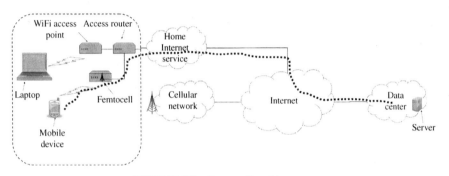

FIGURE 5.7   Femtocell architecture.

A femtocell is essentially a small base station that connects to the Internet access provided that is available. The femtocell can provide connectivity to mobile devices within a small range, typically covering a single home or office building.

Figure 5.7 illustrates how a femtocell will be deployed in a home. The dotted line shows the path taken by the phone when it accesses a server on the Internet. The femtocell connects to the access router and is able to use the available Internet connection to allow the user to communicate. The protocol used by the mobile device to communicate with the femtocell is the same as the one it uses to communicate with the cell tower. However, all such communication bypasses the congested radio access network since the packets are sent over the Internet once they reach the femtocell. This relieves the bandwidth requirement on the radio access network.

5.3.2   Wi-Fi Offload

An alternative to the deployment of the femtocell in the house is to just use the Wi-Fi network as the means to bypass the congested radio network. This approach works well for mobile devices that support network access using the Wi-Fi protocol as well as the cellular protocols. In this case, one need not deploy a femtocell at home but just switch over to the Wi-Fi network whenever one is at home.

If we consider the home network shown in Figure 5.5, a cell phone with both a cellular network interface and a Wi-Fi interface has two alternative options for communication. It can either use the Wi-Fi interface to connect to the Wi-Fi access router at home or it can use the cellular network interface to connect to the cellular network. The Smartphone can be configured so that it preferentially connects to a Wi-Fi network when within its range, using the cellular network only when no Wi-Fi network is accessible. This would then eliminate a significant part of the Smartphone usage from the cellular network every time the phone is used at home when it has a Wi-Fi connection.

The option need not be limited to cell phones at home. There are many Wi-Fi providers who have Wi-Fi hotspots available in many areas, for example, many Internet Service Providers will provide Wi-Fi hotspot service in areas such as airports, train stations, and shopping malls. The Smartphone can be configured so that it preferentially uses the Wi-Fi connection whenever it is in any area where it is connected to an authorized Wi-Fi network which in turn provides connectivity to the Internet.

Wi-Fi offload works effectively for the set of Smartphones that have dual interfaces. Its one limitation is that it would not work for the phones that only support the cellular network interface and requires an additional Wi-Fi interface on the mobile phone. It may or may not be a significant issue depending on the models of mobile devices that are used within the network.

The offloading behavior need not be limited to just Wi-Fi networks but to any other network that may be available and may provide better connectivity for the user. The definition of "better connectivity" in this context may mean a cheaper network, a free network, a faster network, or any network that may have some capacity available.

5.4   BANDWIDTH MANAGEMENT

It is quite possible that all the approaches that have been discussed before for augmenting the bandwidth available to the users of mobile data may still be insufficient to provide adequate bandwidth for all the users on the network. In these cases, bandwidth becomes a scarce resource, and this resource needs to be managed in order to maximize the resulting user experience on a congested network. In this section, we look at the broad category of approaches that can be used for bandwidth management in the radio access network.

The principles that are used for managing the scarce resource of bandwidth are no different than those in managing scarce resources in other conditions.

Nevertheless, the specific nature of bandwidth in radio access network results in nuances in how it is implemented.

5.4.1 Rate Control

Rate control schemes, as described in Chapter 3, can be applied to radio access networks in order to deal with bandwidth shortage. The specific manner in which such a control can be implemented depends on the details of the technology used on the air.

Some of the radio-access protocols like GSM assign a fixed rate to any user who is connected on the network. Other radio-access protocols like LTE and WiMax allow a more dynamic assignment of portions of the bandwidth to the users, and they can be used to restrict users to a specific bandwidth. In the case of the latter, rate control can be implemented by appropriate configuration of the bandwidth limits that a user is allowed to have in the radio network. In the case of the former, any restrictions at the radio level will be hard to enforce, but one can put bandwidth restrictions at an upper layer of the protocol (e.g., at the IP network layer). This would have to be coupled with a scheme that will allow a specific number of users to be actively connected at any single instant in time.

Let us examine the different approaches by which such a rate control on the different users can be applied within a radio access network. In order to explain the approaches, we will model the end-to-end system of the radio access network as shown in Figure 5.8.

Figure 5.8 shows one mobile device accessing some data over the IP protocol. The packets flowing from the mobile device to a server in the Internet need to traverse three segments in the cellular network. The first segment is the link on the air between the mobile device and the equipment at the cell tower. On this segment, the format of the packets that are on the air would look like the diagram shown above the dotted cloud representing the air portion of

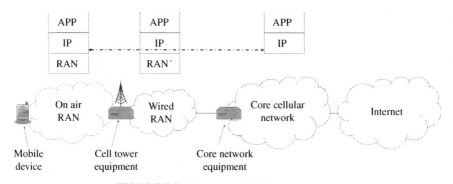

FIGURE 5.8 End-to-end RAN access model.

the RAN. The packet would have a RAN protocol header (the exact nature depends upon the radio network protocol used), followed by an IP header, and followed by application payload (shown as APP in the figure), which may include headers of other protocols such as Transmission Control Protocol (TCP) or Hypertext Transport Protocol (HTTP). The second segment begins at the cell tower. When the network equipment at the cell tower receives this equipment, it would convert them to another set of protocol headers shown as RAN'. The RAN' headers are the ones that will be used to communicate to the core network. Depending on the specific RAN protocol, the RAN's header may itself be an entire suite of other protocols. The third segment begins at the core network. It is the first point where the IP headers are examined, and then packets are forwarded according to the IP protocol specifications. This is shown as the packet with IP at the lowest layer in the figure, even though in reality the MAC layer of some underlying protocol needs to be present for actual transmission. On the reverse direction, an analogous transformation is applied to the different packets.

Rate control to restrict the bandwidth available to a user can be applied at either the IP layer, in one of the upper application layer (APP) protocols, or at the radio network level (either in the RAN protocols that go over the air or the RAN' protocols that are on the wired segment). The rate control can also be exercised differently on the upstream flow (the stream of packets flowing from the mobile device to the Internet) and the downstream flow (the stream of packets flowing from the Internet to the mobile device).

On the upstream flow, the IP layer in the mobile device can restrict the amount of bandwidth that is available to the user of the device. The limits on how much the amount is can come from the equipment at the cell tower or from some other controlling equipment in the radio access network. Alternatively, if the radio access network itself supports bandwidth control, the controls can be implemented at that level.

On the downstream flow, equipment in the core network that processes the IP packets can restrict the amount of bandwidth that is available to the user of a specific device. It may again need to coordinate with some equipment in the access network, which has the knowledge of the current amount to be allocated to individual users. Alternatively, such restrictions on the downstream packets can be done by the equipment at the cell tower if the RAN or RAN' protocols support this capability.

In current networks, it is usual for the downstream link to be more heavily utilized than the upstream link. As a result, mechanisms for controlling the radio network bandwidth are available in the core network using mechanisms such as policy charging and rules function (PCRF).

Policy is an important aspect of managing bandwidth or any other resource that is scarce. The approach of reducing each person to a fraction of its possible bandwidth in the situation of congestion is just one of the many ways in which such an overload situation can be handled. Suppose there are 150 active users when the network is capable of supporting only 100. Instead of giving everyone reduced bandwidth, one might as well decide that only 100 users ought to be accepted into the network at any time and that the remaining 50 ought to be denied access to the network. The difficult decision in that case is to decide which 100 ought to be accepted and which ones ought to be rejected. Should it be strictly on a first come first served basis or on some other basis? Are all users equally important, or are some users more important than others? These are the different policy questions that one needs to decide.

In many cases, the answer to this policy question is that not every user has the same priority. In those cases, different service levels are offered to different sets of users, and the resulting technique is service differentiation.

5.4.2 Service Differentiation

The concepts of service differentiation described in Chapter 3 can also be applied to radio access network. Their application to radio access networks will be analogous to that of rate control.

When the network has more users than can be handled in the available amount of bandwidth, one can divide them into multiple groups of different priority. One can then give preferential treatment to one group over the other. The idea is to preserve the service to the group with the higher priority at the expense of service degradation to the group with a lower priority.

There are many different ways in which such a difference in service quality can be administered in the radio access network. Instead of giving everyone a share of the reduced bandwidth equally when there is an overload situation, one group may get a higher share of the bandwidth than the other. The difference in the bandwidth available allows better service quality to the users who have the higher rate allocation.

An alternative way in which members of a preferred group may be given better service quality is by means of preference during the process of admission control. In all cellular networks, mobile devices need to connect to the network by means of signaling protocols between the mobile device and the equipment at the cell tower (or another location in the cellular network). When there is an overload situation, the requests to join the network are refused to users belonging to the lower priority group, and the users in the higher priority group can connect with a higher probability of success. In essence, it turns out to giving a zero rate to members of the groups with lower priority. In some cases, existing users of a lower priority group may be pre-empted when more users belonging to a higher priority group try to access the network.

The mechanisms to implement the differentiation in service are analogous to the mechanisms used for rate control. The ability to implement all of these controls is available at the IP network layer, and the specifics of the radio network may allow a varying degree of flexibility in implementing these controls at the radio layer.

5.5   NONTECHNICAL APPROACHES

The various mechanisms examined for dealing with bandwidth shortage in the radio access network discussed previously in this chapter are technical approaches. There are also nontechnical approaches that can be used to influence the usage of network bandwidth, and some of the mobile network operators are using those approaches to try to reduce the demand of bandwidth on the radio networks.

The primary driver of bandwidth in the radio networks is the use of data-intensive applications by Smartphones. Until around 2009, most of the wireless network operators in the United States offered unlimited data plans for Smartphones for a flat fee per month. Several operators also offered similar unlimited data plans in Europe. The United States and Europe are the two markets responsible for a large share of Smartphone growth. Since mobile Smartphones allowed users to access a variety of data over the Internet anywhere, the unlimited data plans started to be viewed unfavorably by many of the network operators.

As a result, some operators have started to move towards offering limited data plan or tiered data plans and eliminating unlimited data plans. In these plans, there is an upper limit on how much data a user can consume in a month for a flat fee. If a user exceeds that limit, additional fees are charged depending on the amount of data that exceeds the limit. The rationale for these plans is to persuade each of the users to stay under some limit and to regulate the amount of bandwidth that they consume. The idea is to use higher prices to deter users from using too much bandwidth. Users who consume more resources end up paying more.

The use of such tiered pricing model to shape user bandwidth demand is a double-edged sword. If all the operators in a geographical area offer only limited data plans, one would expect a change in user behavior. However, if a few operators continue to offer unlimited data plans at competitive rates, the subscribers of mobile phone may be tempted to switch to those operators, resulting in a net loss to the operator offering tiered plans. Furthermore, the providers of new mobile devices that are bandwidth hungry may also tend to gravitate towards operators that provide unlimited data plans. Thus, the use of tiered pricing plans needs to be carefully analyzed by an operator to determine if that will be an effective strategy for them.

A variation of the tiered pricing plan does not charge extra for users that exceed the prespecified limits placed on their data plan but starts to apply rate limits if users exceed the limit. While the user can exchange information at the maximum possible rate sustainable in the network till the data usage limit is reached, the rate is reduced substantially if the user exceeds the limits. While it is useful for the users that they do not encounter unexpected high charges on their bill, the issues with such mechanisms are identical to those in the standard tiered pricing plans.

Other operators use other mechanisms to persuade their users to reduce their bandwidth consumptions. These include identifying users who are bandwidth hogs, for example, users that are in the top 1% of the users consuming bandwidth in the network. Disincentives can be put into place for the bandwidth hogs, for example, the rate at which the bandwidth hogs can download additional data is limited once they cross a specific threshold of data per month. The idea being that if such bandwidth hogs switch to a different network, then that would be a better situation for the operator. The approach does need to be balanced against the fact that this is likely to cause customer dissatisfaction. Furthermore, careful consideration must be put into place to determine if such measures should be used on the basis of bandwidth usage alone. If a bandwidth hog happens to be influential in other customer's plans, for example, a head of household with multiple subscriptions to a cellular plan or the owner of a small business with significant influence over a large number of subscriptions, those considerations need to be taken into account before deciding on throttling the bandwidth rates.

The techniques used to detect bandwidth hogs can also be used to detect people who may be tethering their mobile devices. Tethering is the use of mobile device as a way to share the Internet access of the mobile devices with other computers or devices. Tethering includes using the phone as a modem attached to another computer, as well as having the phone act like an access point to which other devices may connect, either via a wired interface or a wireless interface. Tethering can increase the use of mobile data by a device substantially and overload the operator network. Some mobile network operators may have contracts against tethering without the payment of a fee. Nevertheless, there are software packages available that would allow people to use tethering in violation of the contract. Analyzing the network data in real time to detect unauthorized tethering is one of the techniques operators can use to reduce data usage on their network.

Another approach used by operators to alleviate pressure on their networks is to turn off high-bandwidth applications in some cases. As an example, a mobile device may allow video-chat applications that use high bandwidth to work with Wi-Fi connections but turn them off when the network is a cellular one. Restricting the applications that can run on the cellular data network is another nontechnical approach to reduce bandwidth in the radio network.

A similar user-centric approach to persuade people to not use cellular data is to prompt users every time they come within an area that could be serviced by another network that can offload the traffic. Many phones will prompt the users with available Wi-Fi networks every time they are in the coverage area. Reminding users of the availability of alternative modes of connectivity can help reduce some of the bandwidth pressures. Alternatively, phones may be configured to continuously scan for available alternative networks and use offloading automatically whenever possible.

If the bandwidth demand on radio access continues, we are likely to see the emergence of many new technical and nontechnical approaches to persuade people to use less bandwidth.

CHAPTER 6

BANDWIDTH OPTIMIZATION AND COST REDUCTION IN BACKHAUL AND CORE NETWORKS

In this chapter, we look at the specific challenges that a mobile network operator has to face in the area of radio backhaul and core networks. While we have looked at some general approaches in Chapters 3 and 4, this chapter applies those general approaches in the specific context of a backhaul network.

6.1 OVERVIEW OF BACKHAUL AND CORE NETWORKS

As mentioned in Chapter 1, the access network component of the mobile network consists of three components, the radio network over the air, a backhaul network, and a core network. The backhaul network connects the equipment at the cell tower to the equipment that is in the core network. Depending on the specific wireless protocol technology that is used, the function and components used in different elements of the backhaul will be different. However, there are some common underlying technologies that will be used regardless of the wireless protocol that is run atop them.

As mentioned in Chapter 1, networking technologies are designed in a layered architecture, and the layered architecture allows one set of network technologies to be layered on top of another set of networking technologies. The mobile radio network, of which the backhaul is a component, in most

Techniques for Surviving the Mobile Data Explosion, First Edition.
Dinesh Chandra Verma and Paridhi Verma.
© 2014 The Institute of Electrical and Electronics Engineers, Inc. Published 2014 by John Wiley & Sons, Inc.

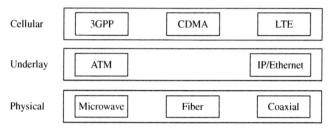

FIGURE 6.1   Layers in the backhaul network.

of the network operators consists of three layers, although one can argue that the bottom two layers should be considered as a single layer. These three layers are shown in Figure 6.1.

The lowest layer is the physical layer, which consists of the links and devices that connect various portions of the backhaul network to each other. The backhaul network could be wired or wireless. In the wireless backhaul, microwave is the predominant technology. In the wired side of the backhaul, many installations used coaxial copper cables to connect cell towers to backend devices. An alternative technology for wired backhaul is fiber optics, which provides significantly higher bandwidth than copper.

The physical layer of the network provides the basic connectivity in the mobile backhaul. The next layer provides an underlay for communication for the cellular protocol. The cellular communication standards specify an underlay of either Internet Protocol (IP) or ATM networks that can be used to implement this layer. IP-based underlay is more commonly used by mobile network operators. In many cases, the IP network would be implemented on top of Ethernet. A special variation on Ethernet, called Carrier Grade Ethernet, is also getting acceptance in deployments. Carrier Grade Ethernet is an enhancement to regular Ethernet, which provides quality of service attributes and can be supported on top of underlying optical fiber or microwave links. More details on Carrier Grade Ethernet are in Section 6.2.

The final layer in the network is the cellular network. This consists of the devices and functions that are specified by the cellular standards. The backhaul in the case of Universal Mobile Telecommunications System (UMTS) networks would typically be the connection between the base stations and Radio Network Controllers (RNCs). In the case of Long Term Evolution (LTE) networks, the backhaul would be the network connecting elements of eNodeB with the Mobility Management Entity (MME). The elements of the cellular network are logically connected to each other using the underlying network for connectivity.

The three-layered architecture is also applicable for the core network of the operator with the exception that the physical layer in core networks would typically be wired networks based on cable or fiber optics, and the use of microwave links at the physical layer is rare.

6.1.1 A Closer Look at Backhaul Technologies

Since the physical network layer is the predominant cause of any bandwidth-related issues in the mobile backhaul, it is worthwhile taking a brief look at the different technologies and their capabilities. As mentioned in the previous section, the three dominant technologies for mobile backhaul are microwave, copper, and fiber.

The mix of technology between microwave, copper, and fiber varies widely between different geographical regions and is highly dependent on the history of how mobile networks got rolled out in the region. In Europe, microwave backhaul is more common with copper taking a second place followed by optical fiber. In the United States, there is less usage of microwave, and wired access is more prevalent. Although copper was the dominant technology for a long time, it has been gradually replaced by optical fiber. In countries like China, the predominant technology for backhaul network is optical fiber. In several countries in Africa, microwave would be the dominant technology.

Microwave-based backhaul has been the dominant method for delivering the backhaul to cellular networks in larger parts of Europe and rural United States. Microwave-based communication happens on selected spectrum that needs to be licensed from the local regulatory agency. Microwave links typically operate in the radio spectrum of 6–38 GHz which can provide speeds of up to 300 Mbps. New microwave spectrums allocated at 80 GHz spectrum (also called the millimeter wave spectrum) allows the speed of microwave to go up to 1 Gbps. Microwave links can be deployed at distances from half a km to 50 km with speeds ranging from 1 Mbps to 1 Gbps, depending on the available spectrum and the nature of terrain. Microwave transmission requires the sender and receiver locations to have a line of sight, but does not require any digging to lay down fibers. This makes this transmission technology attractive as one requiring much less time for installation. On the other side, microwave transmission can be affected adversely by weather and has higher operational costs than wired technologies.

Copper-based backhaul provides connectivity using twisted coaxial copper cable and is generally used to provide T1 (about 1.5 Mbps bandwidth) or E1 (about 2 Mbps bandwidth) line connectivity. They are also frequently used to provide T3 (about 45 Mbps) or E3 (about 34 Mbps) connectivity.

Copper-based backhaul can provide higher bandwidth service. It does require the laying down of cables, like optical fiber. It was the predominant way to provide connectivity to towers before the advent of fiber optics. Although copper is still widely deployed, new upgrades of the backhaul are more likely to go in the direction of optical fibers.

Optical fibers can provide higher bandwidth, which can be up to several Gbps. Optical fibers have virtually no losses, and the use of repeaters allows them to span long distances. Their installation is more expensive, but they provide the nice attribute that once an optical fiber conduit is laid down, additional capacity can be switched on with very little additional expense or delays.

On top of the three technologies of microwave, optics, or cable, one can support different types of interfaces to run the IP or ATM underlay network. These include the specified leased line specifications like T1, E1, T3, or E3, which are standard interfaces for link connectivity. The T series of specifications is used widely in North America, while the E series of interfaces is a European standard. Alternative options include the use of Synchronous Optical Networking interfaces on top of optical fibers.

A promising technology to emerge lately is that of Carrier Grade Ethernet. This technology allows the use of a ubiquitous Ethernet-like interface for the links to connect to the networking equipment, while adding in features such as the concept of a leased line or a tree-shaped topology to connect different elements. Carrier Grade Ethernet also comes with defined quality of service attributes. Wherever Carrier Grade Ethernet is available, it can provide a significant cost advantage compared to other types of technologies.

6.1.2   Causes for Backhaul Bandwidth Limitation

It is worthwhile examining what causes bandwidth limitation in the backhaul. The bandwidth limitation in the backhaul is primarily due to the limitations of the physical layer. It is also determined by the price at which the specific bandwidth is available.

If we look at the wireless physical infrastructure of microwave links, then the limitations in bandwidth come from the physical limits of how much bandwidth the link can carry. In the most commonly deployed microwave technology, the microwave link will provide a capacity of about 34 Mbps. Microwave links at lower capacity can also be obtained at a lower cost. It is possible to use smart ways of sending information and microwave links so that they can go up to 1 Gbps. However, the price of the microwave link at 1 Gbps is much higher than that of the microwave operating at 34 Mbps. Another technical challenge is that upgrading microwave links to higher capacity requires access to additional spectrum on the air, which requires licensing and may not always be available. The cost and technical issues associated with microwave upgrade make it difficult to attain.

The challenges associated with upgrading are significantly less when it comes to the wired infrastructure. Although coaxial cables are usually deployed at T1 lines (1.5 Mbps), they are capable of sustaining a significantly higher throughput and, depending on the modulation schemes, can be used for anywhere between 1 Mbps and 50 Mbps and, with advanced modulation schemes, may even go up to a few Gbps. Fiber optic cables are generally laid out in a group consisting of multiple strands with each strand capable of transmitting up to 40 Gbps or even higher. More than one strands of fiber can be used to get higher speeds.

Despite having the capacity of higher bandwidth possible on the technical side, wired backhaul communication does face a cost limitation even when fiber optics are deployed to the cell tower. If the optical infrastructure is owned by the carrier, it can get additional capacity at little incremental cost. However, if the optical fibers leading up to the cell towers are owned by a different entity, and the mobile network operator has to pay a leasing cost for the capacity of the fiber, the costs of obtaining additional bandwidth to the tower could be fairly high.

Thus, business and cost issues may cause a backhaul capacity to be constrained even if there are no stringent technical issues like that on the air interface.

With this high-level overview of the backhaul, let us look at specific techniques within the backhaul that will help to deal with the bandwidth issues within the backhaul.

6.2   TECHNOLOGY UPGRADE

If the cost is affordable, upgrading the technology of the backhaul network may be the easiest option to address the bottleneck caused by mobile data. The choice of the upgrade technology will depend on the options available at the cell tower and their relative costs. Let us consider the options for each of the technology types.

If the backhaul is a microwave backhaul, then one can upgrade to a microwave backhaul with a higher capacity, transfer the technology over to copper, or upgrade to an optical fiber infrastructure. Each of these options has associated costs, and the lowest cost option can be selected. If the backhaul is copper, then one can switch over to optical fiber. In some cases, it may even be advantageous to switch over to high-speed microwave. When the backhaul is optical fiber, additional fibers can be commissioned to get additional bandwidth.

6.3 TRAFFIC OFFLOAD

Traffic offload is another option for the backhaul. The prerequisite for this option to be exercised is that the cell tower or another point in the backhaul has access to a network that is separate from the one used for normal backhaul communication. If such a network is accessible, one can offload the traffic to that network.

In a rural or remote area, it is likely that the only network available to the cell tower is the one used for the current backhaul, and traffic offload may not be a viable option. However, in urban locations, a single cell-site or areas close to the cell-site may be served by many different networks. In that case, some of the traffic can be diverted off to the other network.

6.4 COMPRESSION

Compression or optimization of traffic is a viable option in the backhaul network. Many of the techniques associated with compression that are described in Chapter 3 can be applied to advantage in the network to save on backhaul capacity.

Given the three-layered structure of the backhaul network, there are a few options for deciding where the compression devices ought to be located. In general, a compression technique requires two devices, one on each side of the link on which bandwidth needs to be saved. On each side of this link, there will be three logical functions that need to be performed as shown in Figure 6.2. The three logical functions are (i) processing required for cellular standards protocol handling, (ii) the transport of data as per the underlay network, and (iii) the transport of data on the physical network.

Compression can be enabled using two logical interception points in the backhaul, one at the intersection between the cellular protocol processing and

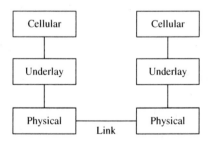

FIGURE 6.2   Logical functions in the backhaul.

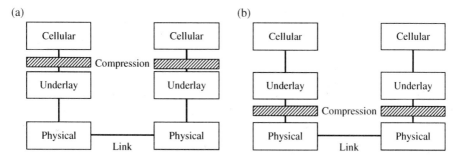

FIGURE 6.3 Options for packet compression in backhaul. (a) The option to compress between the cellular and the underlay (IP) level and (b) the option to compress between underlay (IP) and physical level.

the underlay (e.g., IP-based forwarding) and the other between the underlay and the physical transport. These two options are shown in Figure 6.3.

When done at the latter of these two options (i.e., option (b) in Figure 6.3), packet compression techniques involving IP header compression can be used. In both of these options, other compression such as object compression, general packet compression techniques, and flow sharing and their variants can be successfully used.

6.5 TRANSFORMATION

The transformation of content so that it uses less bandwidth can be successfully used to reduce the bandwidth overload on the backhaul network. An adaptive transformation of the content carried within the traffic can significantly save on the amount of bandwidth used on the backhaul.

Since content is usually carried as application layer payload, one needs to consider the location within the mobile network where the transformation is most easily performed. In order to determine this location, let us examine the manner in which different protocols are layered to carry application-level data in the mobile network. As mentioned in Chapter 1, the IP network visible to mobile devices is an overlay within which the entire cellular network appears as a single hop. That IP network is another layer built on top of the cellular network. As mentioned in the beginning of this chapter, the cellular network past the cell tower consists of three layers. The layering structure of these various systems is as shown in Figure 6.4. The figure also shows the relative positioning of other segments of the network—assuming they are built over other types of technologies as shown in the figure.

Although there is an underlay network, which is IP in many cellular network implementation, this network is hidden underneath the layer of

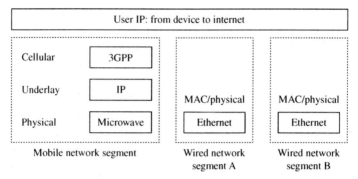

FIGURE 6.4 Layers in core network.

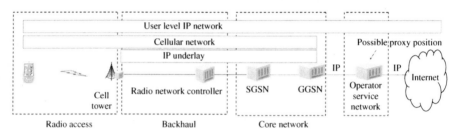

FIGURE 6.5 Proxy location in core network.

cellular protocols and invisible to the user-level IP network that connects the mobile device to the Internet. The application-level objects (e.g., a web-based request for playing a video) are contained in even higher-layer protocols above the user-level IP.

The introduction of a transparent application proxy that can modify the application-level content needs to be done by deploying an application-level proxy in the user-level IP network. Since this user-level IP network is transparent to the backhaul, this network needs to be introduced not into the backhaul, but on the service network belonging to the mobile network operator.

Figure 6.5 shows how the proxy would be introduced for the 3rd Generation Partnership Project (3GPP) networks. The 3GPP network architecture follows a model where the cellular network terminates at a device called the Gateway GPRS Support Node (GGSN). The application-level proxy for content transformation can be deployed in the part of the network between the GGSN and the Internet, which is the place where other types of services offered by the mobile network operator would be located.

The equivalent termination point of the cellular network for LTE networks is called Packet Data Network Gateway (PDN-GW) and that for CDMA networks is called the PDSN (Packet Data Serving Node). The application-level proxy will need to be located between the PDN-GW and Internet, or the PDSN and Internet, depending on the technology being used.

6.6 CACHING

Caching is a technology that can be used for saving the bandwidth on cellular backhaul but it poses some challenges due to the architecture of the cellular networks. Because of the tunneling structure used in the cellular networks, the IP packets originating or destined to a mobile device are hidden inside other layers in the cellular network and only become visible in the core network after the backhaul. Like content transformation services, the natural point for caching would be in the service network of the mobile network operator. However, caching content at that location is not likely to be very effective for saving bandwidth on the backhaul networks since most of the content flows downstream from the Internet to the mobile device, and caching in the service network does not prevent the content from traversing the congested backhaul. Therefore, caching ought to be done at the cell tower in order for it to save congestion in the backhaul.

However, in most cellular network architectures, user-level IP packets (i.e., IP packets sent from the mobile device to a server in the Internet) are not visible at the cell tower. They are embedded within several layers of cellular protocols, since the entire cellular infrastructure consisting of several hops effectively looks like a single LINK/MAC layer for the user-level IP packets. If caching needs to be performed at the cell tower, then user-level IP packets need to be extracted at the cell tower. There are three possible options to perform this type of extraction:

- *First approach*: Perform all the cellular network functions normally done between the cell tower and the termination point for cellular network (i.e. GGSN or PDN-GW) at the cell tower itself. This will allow extracting user-level IP packets at the cell tower, essentially collapsing the cellular architecture to be between the user-device and the cell tower. Extracting user-level IP packets at the cell tower and allowing the implementation of a transparent caching proxy. The transparent proxy responds to mobile device requests for cacheable content, and forwards noncacheable content to the originally intended server.
- *Second approach*: Implement an intermediary which extracts the IP packets from the cellular network and responds to those requests. The intermediary will forward noncacheable content transparently to the originally intended recipient. The intermediary is implemented so that they are transparent to the other devices downstream in the network.

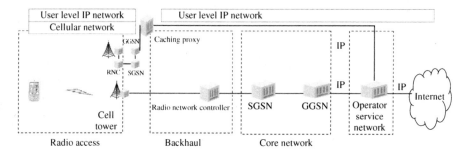

FIGURE 6.6 Caching at the cell tower.

For the case of noncacheable content, the packets flow as they would without the existence of the intermediary. For cacheable content, the responses come from the intermediary.

- *Third approach*: The cellular network infrastructure, which looks like a single LINK/MAC hop to the user-level IP network, in turns deploys an underlay network (usually a hidden IP network, or an ATM network) to run its operations. Instead of using the cellular infrastructure to forward the packets that are not cacheable, the intermediary can forward those packets using the underlay network.

Let us discuss these three models for implementing caching in some more detail. We will use the example of a 3GPP/UMTS network shown in Figure 6.5 to discuss the implementation of caching at the cell tower. Analogues can be developed for other cellular protocols as well. The structure for enabling a caching at the cell tower is shown in Figure 6.6.

In Figure 6.6, the structure of the cellular network as it operates originally is shown in the bottom set of devices in the picture. The components required to perform caching at the cell tower are shown in the top half of the picture. The cellular device functions, namely, RNC, Serving GPRS Support Node (SGSN), and GGSN, are all implemented at the cell tower itself. Generally, one would not find these devices implemented in a form factor suitable for the cell tower, but one can implement software versions of the equivalent capability and put it on a processor at the cell tower. The user-level IP network is now exposed at the cell tower, and caching can be performed like that for any other IP network.

There are a few challenges that one encounters in this model when users move across to other cell towers. Creating an application-level caching proxy usually means terminating the Transmission Control Protocol (TCP) connection that originates from the mobile user. As the mobile user moves to another cell tower, the TCP connection and upper application-level information need to be transferred over to the new cell tower for the caching function to work

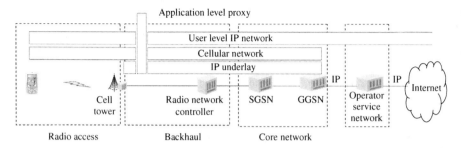

FIGURE 6.7   Underlays with caching at cell tower.

properly. Such a transfer capability can be built into transparent application-level proxies, but the capability needs to be implemented specifically for each application-specific proxy. The other challenge is that the management infrastructure in the mobile network operator may not be designed to support the large number of GGSNs and SGSNs that such an approach would entail.

The second approach to perform caching at the cell tower is more complex, and it involves extracting packets at the cell tower and restoring the stream of packets back into the cellular network subsequently. The process of doing this requires making a transparent proxy that cuts through the different layers of the cellular protocol. The proxy needs to perform all functions required of the cellular protocol, respond back to application-level flows that can be handled and the proxy, and then reinject the noncacheable content into the packet stream. In essence, this approach requires breaking the cellular underlay network into two underlay networks as shown in Figure 6.7 to surface the IP packets at the cell-station.

As shown in Figure 6.7, the system breaks the model of cellular transport in the network, breaking out IP packets and thereby enabling the application-level proxy functions to work. The breakout can happen before the packets are put into the underlying IP overlay, which begins at the cell tower itself.

This type of cross-layer proxy can be created, but is a complex process requiring a close coordination and integration with different layers of protocols. Some of the cellular protocols require encrypted tunnels, and those encryption keys will need to be shared. Cellular protocols keep track of packets transmitted to a user according to their data plans, and the accounting information in most cases happens in devices located in the core network. The accounting information updates would need to be maintained properly for such caching to work properly.

The third option is essentially a hybrid between the first and the second. Once the packets have been taken out at the IP network, one might as well use the underlay IP network to complete the rest of the communication instead of trying to reinject packets in the cellular network paradigm. This approach will allow for a more efficient usage of the backhaul bandwidth since it eliminates some of the layering overhead.

In all of the three techniques for supporting caching proxies at the cell tower, the main challenge of dealing with mobility remains. When a mobile device shifts from one cell tower to another, its connection to the application proxy running at the local cell tower is broken. There are several options to handle this situation:

- Not support mobility for caching proxies: The caching application proxy support is only provided for users or end-devices that are not mobile. In many applications, for example, when wireless network is used for the monitoring of static cameras, or where sensors are connected over the mobile network, the end-point is static and mobility issues can be ignored. Similarly, a large fraction of users do not switch cell tower locations during their sessions, and caching can be enabled for these users. However, this approach may cause dropped connections when users do switch across cell towers.

- Maintain affinity to original cell tower: In any of the three approaches outlined earlier, packets from users that move to a new cell tower can be directed towards their original cell tower. This maintenance of affinity can be done at the IP layer once the user-level IP packets are extracted by using mobile-IP technologies or simple variations on them. At a very coarse level, this approach requires each of the cell towers to keep track of where the mobile users have moved from/to and to tunnel packets to the appropriate location.

- Move application proxies to the new location: An application proxy can follow the client as it moves across the cell tower. A variety of techniques can be used for such a migration. An application-level proxy that is written to support the mobility of the client along with a transfer of TCP session state is one way to address the situation. Another alternative is to implement the applications as virtualized machines running on hypervisors. Virtual machines are developed so as to be able to be transferred across different physical machines. If each user is handled by its own virtual machine, and the underlying infrastructure moves the virtual machines corresponding to a user to the cell tower nearest to it, the application proxies will automatically migrate over to the new location.

Regardless of the approach used, user mobility will introduce overheads, and the usage of application-level proxies is best suited for users that tend to move relatively infrequently. There are sufficiently large numbers of such applications available, which have made it attractive for commercial offerings of a similar nature to appear in the industry [34].

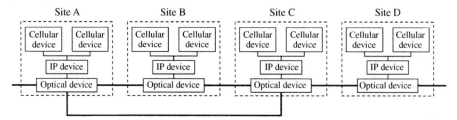

FIGURE 6.8   Structure of mobile backhaul and core sites.

6.7 CONSOLIDATION IN CORE NETWORKS

While bandwidth is an issue in mobile backhaul, it is not as significant an issue in core networks. On the other hand, reducing the cost of operation is always an important aspect in both backhaul and core networks. Given the layering of functionality in the backhaul and core network, the general approach of consolidation can be used effectively in the backhaul and core networks to reduce the cost of operation.

The consolidation approach works well with portions of the backhaul and core network where there is high-speed connectivity. In other words, this cost reduction measure is applicable where bandwidth is plentiful. The idea behind consolidation would be to concentrate most of the functions in a central location. Such a consolidation would generally be cheaper from an operational cost perspective.

Figure 6.8 shows how the general architecture of the mobile backhaul and core network is used to implement the three layers shown in Figure 6.1. Let us consider that portion of the core network that is interconnected using fiber optics technology. In order to implement the three-layered solution, there will be up to three types of networking devices required at each of the location as shown in the figure. At any site, there may be more than one device performing the roles defined by the cellular network specifications (e.g., RNCs of 3GPP system), devices performing the role of the underlay network (e.g., IP routers), and devices performing the role of optical switching and transmission. The different locations will be connected together by high-speed optical links that are shown using bold lines in the figure. The number of each type of device that is present depends on the specific architecture of the network element that has been rolled out. The links that are leading out of the left-most and right-most sites indicate the connections to other types of networks (e.g., to equipment at the cell tower for air network interfaces, or to external networks).

Since the optical links provide high-bandwidth and low-latency communication between the different sites, the network function can be consolidated into a single location as shown in Figure 6.9.

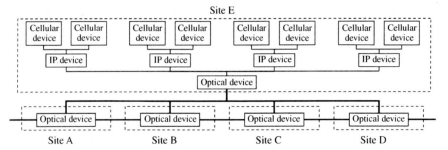

FIGURE 6.9   Structure of consolidated backhaul and core sites.

In this revised architecture, each of the various distributed sites can be made very small, and all the other devices can be consolidated into a single site. In Figure 6.9, we have shown the consolidation occurring at a fifth site—site E. However, it could be one of the existing sites that could be expanded while reducing the footprint of the other sites. Concentrating the different devices into a single location can also help in reducing the overall number of devices that are required in total across the environment.

The amount of cost saving that can be realized from the consolidation of the different sites depends on the nature of the network architecture. However, in most of the cases, reducing the complexity of several sites will result in a significant reduction in the operational and capital expense.

6.8   NETWORK FUNCTION VIRTUALIZATION

Network function virtualization (NFV) is an initiative in the telecommunication industry led by a group of leading network operators [35]. They have come together to form a working group in the standard body European Telecommunications Standards Institute to define the requirements of NFV. The group started operation in January 2013. The primary focus of this group can be seen as an application of consolidation in the core.

Currently, telecommunications networks are built from many distinct networking devices, each device performing a dedicated function. NFV seeks to disrupt this paradigm and moves towards a new network architecture where the network functions are delivered by software running on standard IT servers, essentially as virtual appliances.

As an example, currently, devices that provide the functions of MME/ SGSN as defined in LTE specifications are available as special function hardware appliances from the leading network equipment providers. They are based on specialized network processors that provide the required capability. In the vision of NFV, equivalent capability can be provided as software running on a standard IT server.

At a high level, NFV can be viewed as implementing the function provided by a network device along the following stages of increasing virtualization.

Stage 0: This is the current stage where the network function is delivered as a dedicated appliance built from special-purpose hardware with specialized components. These dedicated appliances are distributed at various locations in the network.

Stage 1: In this stage, the network function is delivered as a dedicated appliance, but the internal structure of the dedicated appliance consists of the function implemented in software running on an IT server. The software itself may be implemented as a virtual machine on the server or as a tightly integrated bundle of software on servers for improved performance. This changes the internal structure of the network device, but the distribution of the devices in the network remains the same. Some of the equipment offered by network equipment providers already implements this model.

Stage 2: In this stage, the network function is delivered from software (either virtualized images or implemented natively on the underlying hardware) that is running on servers in a centralized IT system. The locations where the appliances were placed contain a fast Ethernet switch (or a fast optical switch) that carries data rapidly to the centralized IT system where network functions are implemented in software. Stage 2, in essence, centralizes the distribution of appliances to a few locations within the network.

Stage 3: In stage 3, the software implementation of different appliances is delivered from a cloud-like virtualized environment where network functions are implemented as virtualized appliances. In the same way that an IT cloud adjusts its load among different applications, the networking functions will be supported on a cloud infrastructure.

The attainment of the final stage, where network functions are implemented as software running in a cloud server, will offer the greatest savings. The ability to reach this stage depends on the amount of compute capacity that is available in the cloud and the total cost of operating the consolidated infrastructure. The cost of computing is falling faster than the cost of network equipment, driven primarily by the economies of scale, given the fact that computing devices are used in much larger volumes than networking equipment.

In addition to a reduction in cost, a cloud-oriented network implementation offers the ability to create new networking capabilities rapidly, since it can be created by changing software in a single cloud location instead of rolling out equipment at many different locations. In general, creating centralized applications is easier than creating distributed applications. Because of cost as well as flexibility benefits, it is quite possible that the cellular core network infrastructure a decade from the printing of the book would consist of network capability consolidated in this manner, being provided using the techniques of NFV.

6.9 COST REDUCTION OF THE SUPPORTING INFRASTRUCTURE

The approaches considered till now in this chapter focus on reducing the cost of the data transfer infrastructure in the core and backhaul network. In addition to the data transfer infrastructure, there is supporting infrastructure that is present in the core network of most operators. The supporting infrastructure includes the tools that are deployed in the network for network management functions, such as network monitoring, performance management, security functions, billing and charging infrastructure, and tools for root cause analysis and trouble shooting. Most of these operations are performed by inspecting the contents of the packets that are flowing through the network. Security monitoring looks for suspicious patterns within the network packets, while performance monitoring keeps track of the number of packets and bytes transferred in the network, etc. As the amount of mobile data grows on the network, the cost of the supporting infrastructure will also increase. Although the support infrastructure is usually less expensive than the data transfer infrastructure, it would be beneficial to be able to reduce its costs as well.

Reduction in the cost of support infrastructure may be achieved by means of increasing automation and processing within this infrastructure. A large part of the cost of the support infrastructure is manual. Network performance problems or network failures need to be detected and diagnosed by network administrators. As the volume of data grows, such problems tend to increase more, requiring more automation. Techniques that can automate as much of the manual process are useful in reducing the cost of operation. An automated system, which can use a predefined set of rules or logic, to perform the most common and repetitive network management functions, can reduce the cost of network operation significantly.

Another supporting infrastructure that is expensive in core network is the customer help desk. As mobile data explodes, and new models of mobile devices are included in the network, calls to the customer help desk keep on increasing. This effectively increases the cost of operations. Providing self-help ability to customer, for example, using a web-based portal for them to perform basic operations on the accounts, alleviates much of the need for expansion. Similarly, providing customer forums where customers effectively solve each other's problems can reduce the cost of support desk infrastructure significantly.

CHAPTER 7

CONSUMER-ORIENTED DATA MONETIZATION SERVICES

Three important challenges for the mobile network operator are (i) how to get the maximum bandwidth out of its existing network that may be getting congested, (ii) how to reduce its cost of operations as the data growth happens, and (iii) how to make more money and profit from the data flowing on the network. The first two of these challenges have been addressed in the previous four chapters. In this chapter, we examine the techniques available to the network operators to make money off the data that is flowing on their networks, or in other words, monetize the data flow.

As a rule, mobile network operators can monetize the data flowing through their networks by offering some services for which other members in the mobile data ecosystem described in Chapter 2 are willing to pay additional money. The two primary ecosystem groups from which the mobile network operator can obtain additional money are its users and the application service providers. The users of a mobile network operator fall into two categories: consumers and enterprises. The mobile network operators can provide additional services based on data networking to both types of users to make additional money. Similarly, it can offer additional services to other application service providers.

In this chapter, we examine some of the services that a mobile network operator can offer to consumers, that is, people who have purchased a data plan from it. These are new mobile data–oriented services that will be attractive

Techniques for Surviving the Mobile Data Explosion, First Edition.
Dinesh Chandra Verma and Paridhi Verma.
© 2014 The Institute of Electrical and Electronics Engineers, Inc. Published 2014 by John Wiley & Sons, Inc.

enough for some of the consumers to pay an extra amount of money to mobile network operators in order to avail of these services.

In the telecommunications terminology, these services would normally be classified as value-added services. However, there is a significant difference between the value-added services that are described in this chapter and traditional value-added services provided by the operators. The services that are provided by the network operator in this chapter are derived from the content and properties of data flowing on the mobile network, in contrast to traditional value-added services that are focused around cell phone calls, Short Message Service (SMS)-based texts, and related enhancements.

The ways in which these new data-oriented services can be offered to the customers of mobile network operators are similar to traditional value-added services. These may be offered for *free*, as a special service that helps to retain customers, for a *premium* for which a fee is charged for the service, or as a *freemium*, that is, a service that is offered for free but sells updates or modifications for a charge.

The relationship between mobile network operators and application service providers is that of cooperative competition. In many cases, the services that a mobile network operator can provide to its users can also be provided by an application service provider. The mobile network operators refer to these services as over-the-top (OTT) services. Depending on the nature of the service, either the mobile network operator or the OTT service provider may have an advantageous position in offering that service. The properties or attributes of the mobile network operator that give it an advantage in offering a service are its differentiators. Before describing the new services, it is important to examine the differentiators that mobile network operators as a group may have over OTT providers as a group.

7.1   MOBILE NETWORK OPERATOR DIFFERENTIATORS FOR CONSUMER SERVICES

A differentiator is a property that the mobile network operator can provide to its consumers that others providing the same service do not have. As services are offered by the mobile network operators to consumers, these properties would tend to encourage the consumer in getting the service from the mobile network operator, rather than some other provider.

One of the differentiators that the mobile network operator has is an existing relationship with the consumer. The existing relationship includes aspects of billing and trust. The consumer already receives a bill from the mobile network operator, and it will be more convenient for the consumer to buy another service from the mobile network operator if it is billed together.

By entering into the relationship, the consumer has already established a trusted relationship with the mobile network operator. The consumer has to put a level of trust in the mobile network operator that the services provided by them will meet some minimum level of quality, that it is safe conducting business with them, and that the mobile network operator will not abuse the information it has about the consumer.

This existing relationship can provide an advantage to the mobile network operator when offering services to consumers. Some of the services that the mobile network operator can offer to the subscriber are listed subsequently.

7.2   SINGLE SIGN-ON SERVICE

One of the nuisances in accessing any service over the Internet is the need to maintain passwords for many different Internet sites. There are user identity and passwords for almost every site a user visits, and many of these sites have different rules for creating and maintaining passwords. Some of the sites would require a user to use a mix of capital and small letters, others would require at least some numbers to be used, while another one may require the use of nonalphanumeric characters like & or ! to be a part of the password. Depending on the security policies of the website, they may also require the passwords to be changed on a periodic basis or mandate that a user access the site at least once every few months in order not to deactivate the account. In general, managing all of the passwords for different sites is a painful and onerous task. If one could have a single user identity and password for different sites in the Internet, the life of many users will be very convenient. That will provide all users with a single way to sign onto the various websites.

The problem is significant enough that several solutions and approaches to address the single sign-on problem have already been proposed. In general, there are two broad approaches for providing a single sign-on solution, one based on trusted credential providers and the other based on trusted proxies.

The general system setup for single sign-on based on trusted credential providers is shown in Figure 7.1. In this solution, there are three types of entities: a user, a credential provider, and one or more resource providers. Note that we are showing the user on a mobile device in the figure in accordance with the scope of the book. However, most single sign-on solutions are designed for a general space where the user can also be on a different type of device, for example, a laptop or a PC connected to the Internet via a noncellular network. The user will use the credentials from the single credential provider to access services of the resource provider. Instead of using different credentials for each of the resource providers, the same credentials, the one

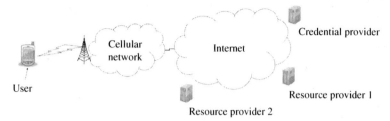

FIGURE 7.1   Internet-based credential providers.

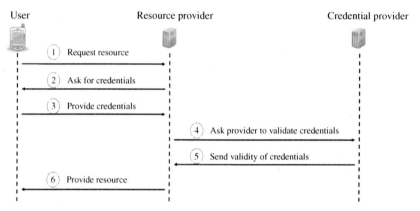

FIGURE 7.2   Steps in authentication and access.

issued by the credential provider, will be used to access the information at each of the resource providers.

Figure 7.2 shows one possible sequence of steps involved in authenticating and accessing resources at the resource provider by the user. These steps assume that the user has already obtained appropriate credentials from the credential provider. In the first step shown in the process, the user requests a resource from the resource provider. As a second step, the provider asks the user to provide their credentials. The user responds with the credentials previously provided by the credential provider. In the next step, the resource provider sends the information to the credential provider in order to validate the credentials. When the credentials are validated, access to the resource is provided.

There are variations in the sequence of steps that can be used when the resource provider is not trusted. An alternative sequence is shown in Figure 7.3. In this sequence, the user does not provide the credentials to the resource provider directly. Instead, the resource provider directs the user to obtain an authorization token from the credential provider. The user obtains the authorization token from the credential provider, which would be valid for a limited amount of time. The authorization token is presented to the

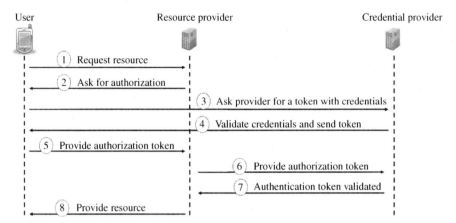

FIGURE 7.3   Alternative steps in authentication and access.

resource provider, who can validate it from the credential provider. In this sequence, the raw credentials are never exposed to the resource providers. Other variations of the sequence, for example, obtaining the authentication token before asking for the resource in the first place, can also be used.

There are various implementations that support authentication by means of a trusted credential provider. The Kerberos system [36] and the popular OpenID [37] scheme used on the Internet are examples of credential provider–based systems. Some companies on the Internet act as credential providers for OpenID-based credentials. Popular social network sites or popular email sites on the Internet can make for good credential providers. Suppose a social networking site *SN* is acting as a credential provider for OpenID. Using this scheme, web surfers can use their login at *SN* website to access services at a site other than the credential provider, for example, at a news site *NW*. *NW* would not need to have its own independent credential management system and can rely on the credentials of *SN*. In the terminology of mobile network operators, credential provider *SN* is providing an OTT solution for single sign-on.

In the proxy-based system, the credential provider is an intermediary in the flow between the client and the resource provider. The relative location of the intermediary and the resource provider has to be as shown in Figure 7.4. The intermediary has to be placed in the network such that the flows between the user and the resource providers always traverse through the intermediary.

In order to provide the single sign-on function, the credential provider intercepts the requests flowing from the client to the resource provider, validates the credentials from the user, and then includes the credentials for that resource provider in the requests sent to the provider. The credential provider could use a different set of credentials when it forwards the requests from the user to the

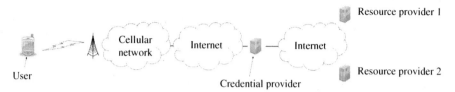

FIGURE 7.4   MNO as credential provider.

User	Credential provider	Resource provider

① Request resource

② Ask for credentials

③ Provide credentials

Credential database

④ Ask for resource

⑤ Ask for credentials

⑥ Look up database provide new credentials

⑧ Relay resource   ⑦ Provide resource

FIGURE 7.5   Steps in authentication and access with MNO as credential provider.

resource provider than the ones it provides to the user. The credential provider can also store a different set of credentials for different resource providers in the system, which can be retrieved from a database whose access is provided by means of security credentials between the client and the credential provider. The flows for authenticated resource access in such a system will be as shown in Figure 7.5.

The proxy-based single sign-on systems are easier to implement using the standard Hypertext Transport Protocol (HTTP) protocols since one can easily insert credentials into the HTTP protocol request streams. When an encrypted protocol like HTTPS is used, URLs can be rewritten so that they create an HTTPS session from the user to the intermediary and a separate HTTPS session from the intermediary to the resource provider. This type of manipulation, as well as inserting credentials, requires a high degree of trust in the intermediary service by the user.

Despite the adoption of OpenID by many sites on the Internet, single sign-on remains an issue for users of the Internet. In order for the credentials provided by a site such as *SN* to be used on another site such as *NW*, an explicit arrangement between *SN* and *NW* needs to be made. In general, the resource provider needs to have an agreement with the identity provider whose credentials are accepted. Every website on the Internet is not likely to accept *SN* credentials, especially websites that are owned by companies competing against *SN* even if *SN* happens to be used by a large number of people on the Internet. The ability to use a single identity is limited to those sites that have agreed and made arrangements to accept those identities.

Another issue with the OTT solution is that there are many providers of credentials on the Internet. There are many popular website operators on the Internet who all provide their own set of credentials. The user needs to decide which of the many credential providers it will use. If one takes into account all the shopping sites, email accounts, bank accounts, school and educational resources accounts, news sites, and entertainment sites that a typical user visits in the course of a month, the number of websites requiring credentials can easily run into hundreds. If a user tends to access a hundred sites relatively frequently, even if those hundred sites use the credentials provided by six credential providers, the user is required to remember and use six sets of credentials to access the sites it needs to authenticate itself to. The problem of remembering six set of credentials is easier than the problem of remembering a hundred set of credentials, but it is still not as attractive as just using a single set of credentials.

The other challenge involved in using a trusted OTT credential provider is that a set of complex interactions need to be supported in order for the credentials provided by the identity provider to be used by the resource provider. The set of network exchanges that would happen normally between a browser and *SN* are completely independent of the network exchanges that happen between the browser and *NW*. To link these interactions, a combination of methods like cross-site scripting, embedding links to another site within one site, and storing specific cookies in browsers needs to be used. Some of the methods may not work if users configure their browsers to eliminate storing browser cookies or eliminate features like cross-site scripting due to security concerns. These complex interactions can also pose security vulnerabilities [38], though they can be largely addressed with proper and careful implementation on part of the identity provider and the resource provider.

Another major challenge in having an OTT service using secure credentials comes in the area of trust. Let us assume that the credentials provided by a site like *SN* can be used to log on to the sites of a hundred other websites that a user frequents. Even though the user may use *SN* to remain in touch with a circle of friends, he or she may not want to trust knowledge of all the history of web browsing to be accessible to *SN*. The lack of trust may be well founded, because the user may not like the privacy policies offered by a credential provider like *SN*, or it may be unfounded based on rumors, emotions, or irrational logic. Regardless of the justification or lack thereof in trusting a credential provider to know about Internet access history, some users are uncomfortable in trusting a single site to know about all of their web-based activities.

Because of the issues involved with ubiquity, security, and trust relationships, mobile network operators have an advantageous position for their subscribers when it comes to providing single sign-on services. In the case of mobile network operators, all network exchanges from their mobile users happen to go through their networks, and this allows them natural points to intercept their requests. This allows for a mobile network operator to provide an easy intermediary-based mechanism for single sign-on.

The single sign-on service provided by the mobile network operator is coming from an entity that is already trusted by the subscriber. The mobile network operator is trusted to carry the packets that are being carried between the mobile user and any server on the Internet. The mobile network operator already has access to the entire web-browsing patterns of its subscribers. Asking it to manage the credential store does not provide mobile network operator any more information than what it already has.

There is some additional information that the mobile network operator would need to have from subscribers when it provides this value-added service that it does not have currently. It will need to have access to and manage the credentials of the user with various websites. The user will have to trust the mobile network operator with managing all of the passwords in a secure and efficient manner. The liability of providing this trusted and secure store is the main risk involved in providing this value-added service. Some of the risks may be reduced by encrypting the contents of the secure store and splitting the encrypted content into multiple portions, each portion being stored in a separate server. Even if the security of one server is compromised, for example, a disgruntled employee reveals the contents of the server on the public Internet, the information cannot be decrypted without the content on all of the other servers, thereby mitigating the risk of revealing sensitive information.

If the mobile network operator provides the sign-on mechanism as a proxy for browser protocols, for example, as a web proxy for all applications that use web-based access, the mobile network operator sign-on system will work for any website without requiring any agreements in place between the operator and the website owner. Thus, the single sign-on service provided by the mobile network operator can be much more seamless for the subscriber to use than any OTT alternative.

7.3 PRIVACY SERVICE

Some users in the mobile Internet are worried about their privacy and anonymity. They would prefer not to have websites they visit track their requests, via browser cookies or via recording the Internet Protocol (IP) addresses that they deploy. The reasons for desiring anonymity and privacy differ from individual to individual, one may desire it because they are dissidents in

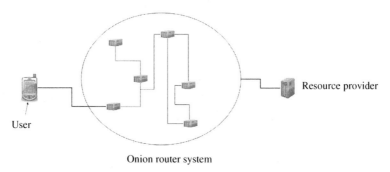

Onion router system

FIGURE 7.6   Onion routing architecture.

a repressive regime, another may desire it to avoid being targeted by attempts to cross-sell unnecessary items by sites being accessed, while a third one may desire privacy because of unwarranted paranoia. Regardless of the validity or rationale for desiring privacy, it can be safely asserted that there would be a segment of people who would desire anonymity and privacy.

The current state-of-the-art technology for achieving privacy over the Internet is using the technology of onion routing [39]. The basic concept is shown in Figure 7.6.

Instead of connecting directly to the site on the Internet, the user interested in preserving its anonymity would connect to it via a set of intermediaries. The request will be passed randomly among the intermediaries, and finally one of the intermediaries will connect to the desired site. The site only knows the identity of the intermediary that it directly connects to and does not know the identity of the original client. Thus, the identity of the original client is preserved.

Furthermore, in order to not rely on the security characteristics of any of the intermediaries, the client would use a series of encryption schemes to route the packets among the set of intermediaries. It would randomly select a route among the different intermediaries. It would then perform a number of encryptions on the content that is to be forwarded, with each level of encryption containing a set of instructions to forward the packet to the last level. The final level of the encryption would indicate the site to which the message needs to be forwarded.

Let us consider the situation where the client C has decided to access a site S on the Internet using the path among intermediaries 1 through 4 as shown in Figure 7.7. The client C wants to send the information shown as Data in the figure to site S, but wants to hide its identity from S. The client C creates a packet that will encrypt the data such that only intermediary 4 can decrypt and extract the data sent to it. One way to do it will be via public key cryptography, for example, by encrypting the data contents with the public key of

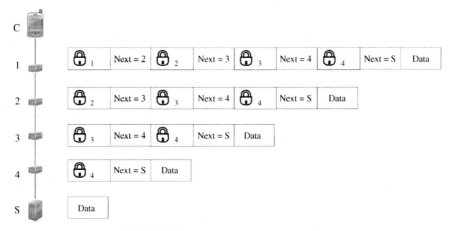

FIGURE 7.7   Onion routing packet headers.

intermediary 4. In Figure 7.7, it is shown by a lock marked 4 in front of the data. It would then attach the direction that the next hop ought to be 4 and encrypt that information so that only intermediary 3 can decrypt that information. This is shown as the lock marked 3 in the figure. An instruction that next hop should be 3 is included and the aggregate locked so that only intermediary 2 can decrypt it. Similarly, the instruction that next hop ought to be 2 is added and then encrypted so that only intermediary 1 could decrypt it. The result is the packet structure as shown at the top right column of the figure. As this message is sent out, each of the intermediaries can decrypt the content of the packets received to determine the next hop and relay the data. At each hop, each of the intermediaries can store the identity of the previous intermediary and relay the information back in a hop-by-hop manner without knowing the identity of the original sender. On the Internet, there are services available that support onion routing technologies, the most popular among which is the Tor Project [39].

When used in the context of web surfing, the concept of onion routing can hide the identity of a web surfer from a website and vice versa. However, the identity of the web surfer is known to the first intermediary that it connects to. At the very least, the IP address of the web surfer is known to the first intermediary. Similarly, the data being sent to the website and the identity of the website are both known to the last intermediary. Thus, the anonymity function requires a trust relationship on part of the web surfer or website to the set of proxy servers providing onion routing. Another issue with onion routing model is that the series of encryptions required to transfer the route adds a significant amount of latency and inefficiency in communication.

Since the mobile network operator has to know the IP address of its subscribers, exposing the IP address to the operator does not reveal any information that the operator already does not have. If the mobile network operator were to offer an anonymization or privacy service, it would have an advantage compared to the same service offered by an unknown provider on the Internet. It already would have the trust relationship that the subscriber would need to have with the OTT service provider. Furthermore, the operator can also provide the anonymization in a much more efficient manner.

The only case where this trust relationship would not work is when the subscriber is trying to hide its identity from the local government or regulatory agency. In almost all jurisdictions, a mobile network operator would be required to give up information to the government or regulatory agencies when requested in accordance with prevailing regulations. Therefore, if anonymity is desired from the government, the mobile network operator would not be the preferred provider of this service.

For the other larger user population, who are interested in anonymity or privacy to prevent other sites from tracking their behavior, an anonymity service from the mobile network operator could be very attractive.

7.4 CONTENT CUSTOMIZATION SERVICES

Content that is available on the Internet may not always be available in the format that is most consumable or attractive to an individual browsing the Internet. The following is an enumeration of some of the few ways in which a website may be misaligned with the needs of an individual trying to access it on a mobile phone.

- The user may have a small screen, while the website was designed for a laptop user with a big screen.
- The user may have limited bandwidth on the air, while the website contains videos that are encoded in high definition that overwhelms the available bandwidth to the user.
- The user may be more comfortable in a language different than that of the website.

In order to address any of these incompatibilities, one can go through an intermediary proxy that provides the function of matching the capabilities of the user to the capabilities of the website. This matching function may require translating the website contents or adapting the website content so that it matches the requirements of the user better.

Such an intermediary proxy can be provided on the Internet, or it can be provided by the mobile network operator. If a company were to offer such a service on the Internet, it will need to figure out how to get requests from the users to any website routed to the company's servers. Let us look at the steps involved in this process by which a service on the Internet can offer this feature. For the purpose of illustration, let us assume that the service is a translation service that would translate any page to Korean from the original language for a person who prefers to see everything in Korean. In order to surf the web in a manner such that the entire contents appear to be in Korean, the user would need to make sure that the Korean translation page is always in the path of any communication between the user and translation web page. The following are the choices available to the Korean translation website in order to enable this to happen:

- It could instruct the users to configure their browsers so that the Korean translation website is set to be the next-hop proxy for all communication.
- It could provide its own private version of a browser, or a browser plug-in that translates any content into Korean when it is downloaded.
- It could instruct the users to access other sites via first accessing the Korean translation website, provide a way for them to input the location they want to surf, and rewrite embedded links on any pages that are accessed so that any subsequent accesses go through its site.

Each of these options has its drawbacks. Users are not always comfortable changing their proxy configurations. Furthermore, the situation can be difficult to manage when a user wants to access two content transformation services, one dealing with Korean translation and the other dealing with converting content to work on smaller screen sizes since the browsers can only be configured to go to one of these two services. Providing a private version of the browser or a customized browser plug-in requires supporting the different types of browsers and their versions that different users might have. And customizing the contents of different pages to alter their links is also a complex process. If the pages are all linked via simple links, it would be easy to do. However, since some of the websites use complex scripts to determine which pages and links to download, ensuring that all the links are rewritten properly requires some sophisticated analysis of the content included within the scripts—which itself may be written in a variety of languages such as JavaScript™ or VBScript, or other lesser popular languages invoked by means of browser plug-in and extensions. While none of these problems are insurmountable, they do make the task of a website trying to customize content difficult.

  In contrast, the mobile network operator enjoys a natural advantage when providing such type of customization for their subscribers. The subscriber requests and responses are naturally flowing through their networks and provide convenient points where a transparent proxy for any communication protocol can be introduced. As a transparent proxy, the mobile network operator could easily fall back to the property of the HTTP protocol that all content flows through a request–response pattern and could provide any customization on a request-by-request basis.

In almost all situations where an intermediary- or proxy-based customization is required, the mobile network operator will have a logistic advantage over other types of services that offer equivalent customization.

7.5 LOCATION-BASED SERVICES

The mobile network operator has information about the location of the different subscribers on its network. It can use that information to deliver services that may be beneficial to a subscriber.

A location-based service is a service that is offered to a user based on the user's location. The service typically uses the information about a user's current location, but can also be based on past or present locations of the users. In general, information about locations can be used to create services oriented towards individual consumers, or services oriented towards enterprises and businesses. In this chapter, we will focus on the services that can be offered to a consumer. The next chapter discusses flavors of location-based services that can be offered to enterprises or businesses.

The following are some examples of location-based services that can be offered to consumers. Note that not all of these services provide a good opportunity for data monetization for mobile network operators.

- Providing information about the nearest facility, for example, a restaurant, parking garage, bank branch, or ATM location.
- Providing information about the possible route to reach an address or a destination.
- Providing alerts about any possible hazards in the area.
- Providing information about friends or colleagues who may be present in the same neighborhood.
- Providing the information to a user about the location of a lost cell phone.
- Allowing parents to track the current location of their child.

From the perspective of how location-based services are provided, these services can be classified into two categories—push services and pull services. A location-based service is a pull service if it is requested explicitly by a user or software on a user's cell phone where it provides its location information. A service in which a user can look up the location of the restaurants within a quarter-mile radius would be a pull service. A push service is a service in which information is being sent to the user without the user requesting it immediately in an explicit manner. An example would be a notification about any possible hazards that exist around the current location of the user. A push-based location service could be delivered as an SMS or Multimedia Messaging Service text message to a user's cell phone, or a visual/audible notification pop-up that may come up on a Smartphone. A user may preregister its interest in receiving specific types of push notification in advance, and the system could provide the information when it is appropriate.

Location-based services can be provided by a mobile network operator or by another service provider that is running over the Internet. A mobile network operator has access to the locations of all of its subscribers. However, with modern Smartphones that have GPS capability, Internet-based service providers can also get access to the location of users that are accessing their services. The mobile network operator has some advantages over such service providers. It can determine the location of its users even if they do not have a Smartphone, which may be true of many subscribers in the developing world. With the current state of the technology, turning on GPS capabilities in Smartphones has a non-trivial impact on the battery performance. In contrast, a mobile network operator can compute position based on information available in the cell towers without requiring any computation on the cell phone itself. Furthermore, with the exception of a few very large Internet-based service providers, a mobile network operator has access to the location information of a much larger set of users. In order to offer a service that will be more appealing to the customers, the mobile network operator has to look at location-based services that have a key dependency on one of the advantages it enjoys.

One such location-based service where the mobile network operator may have an advantage is in the case of emergency notifications. If a user happens to be in an area where there is some danger, for example, a downed electric wire, or a potential threat of bad weather, it will be useful to get a notification of the danger so that appropriate planning action can be implemented. Offering such a service requires constant monitoring of a user's location, determining if the user is in a situation where an emergency notification is required, and then sending an appropriate notification, for example, an SMS text or another type of alert. The mobile network operator has the advantage that it knows the location of any of its users regardless of the type of phone they are using. In order to stay connected to the network, each cell phone needs to be connected to at least one of the cell towers in an area. In many cellular network protocols, the cell phone is associated to more than one cell tower, even if one of them is the primary path for communication. Thus, the mobile network has the ability to tap into its network information and determine the location of a phone without imposing any overhead on the cell phone.

In contrast, location determination by an OTT provider requires some software running on the cell phone. This software needs to monitor the cell phone's current location, report it to a server in the Internet, which can then advise it for information relevant to it. For an emergency notification situation, which may be applicable in a very small set of circumstances, there is a lot of overhead and unnecessary communication with adverse effect on the battery life of the phone. Furthermore, if the data plan of the user with the provider has a limitation on the amount of bytes transferred or a per-byte cost associated with data transfer, such communication may come at a cost to the phone user. The mobile network operator, even if it were requiring data transfer for any of its services, can waive charges for such communication.

Because of these factors, an emergency notification service provided by an operator may appeal much more to a subscriber.

These emergency notifications need to be pushed to the user, which means a service located in the network (either the operator's network or the Internet) has to proactively send information to the cell phone user. This is in contrast to a pull-oriented location service, a service in which a user requests some information from a service providing its information as an input. An example of a pull-based service would be a request from a user to find out restaurants around an area where he or she is currently present. In a pull-oriented location service, the mobile network operator does not enjoy much of a technical advantage when compared to an OTT service provider. It can exploit a non-technical business advantage it has, namely, offer data communication to its own service for free. The relative significance of this advantage depends on how much data exchange is required for offering the service. In many services, the data exchange may not be too high, which reduces the advantage the mobile network operator may have.

In contrast, the mobile network operator has some substantial advantages in offering push-based location services, that is, service in which a notification needs to go from the service to the consumer. The emergency notification service is one of these services. Other examples of push-based services include the following:

- Notification of neighboring facilities: A user may notify the service that it is interested in finding parking for his car when driving. The service can notify the user of any parking lots with available spots that are nearby as the user drives through an area.
- Notification of congestion: A user may be notified of congestion along the path that he or she is traveling on, or in the general neighborhood of the user's current location. This would help the user avoid congestion in the area and provide more localized congestion information than the general area.

- Notification of friends or family members in the area: A service may track and notify users of other users (designated as friends of the users within the service) if they happen to be within a certain proximity range of the user. Privacy concerns need to be taken into account when offering such a service.

- Notification of special events in the neighborhood: A service may track the position of a user and notify him or her of any special events in the neighborhood, for example, a political rally or parade planned in that area or planned street closings in the neighborhood. Such notifications may help the user in deciding how to manage the impact of these events or to attend the event depending on their preferences.

There are many more push-based location services that are likely to come up as a result of innovative ideas from service providers.

7.6 PHONE-BASED COMMERCE

One of the unanticipated results of the popularity of cell phones in regions such as Africa and Asia has been their use to conduct banking at small scale or allow microfinance. This has led to creation of services such as M-PESA [40]. This service (M for mobile, PESA is Swahili for money), offered by a mobile network operator, allows consumers to use their cell phones to make deposits, withdrawals, transfer credits to other users, and buy airtime minutes using their cell phones. In some areas, the service can also be linked to bank accounts.

The mobile network operator has the ability to use its relationship with the consumer to offer a commerce-oriented service to its subscribers. There are different regulations on activities that deal with managing money in different countries. However, in most countries, a mobile network operator can use the mobile minutes and mobile data plans as commodities that can be traded among the different subscribers within the scope of regulations.

This allows the mobile network operator to offer services that can permit the transfer of minutes from one account holder to another. Many mobile network operators allow phones belonging to a single account or a family account to share plans. They can also allow unrelated accounts to transfer minutes to each other charging a small commission in the process.

The mobile network operator can provide applications that provide the concept of a mobile wallet. This works like a prepaid account against which expenses can be charged by participating outlets and retailers. The concept of a mobile wallet brings mobile network operators within the realm of banking operations, and the regulations of different jurisdictions ought to be taken into account when offering these services.

Mobile commerce services can be offered with or without the introduction of new technologies like Near Field Communications (NFC). NFC technology allows for a more secure method for authenticating cell phone users as they are making payments at terminals. However, depending on the business risk and the regulatory environment prevailing in a county, NFC may or may not add a significant advantage to providing mobile commerce.

7.7   OTHER SERVICES

The previous sections listed some of the services that a mobile network operator can offer to its consumers. However, a mobile network operator can also offer any service that an application service provider can offer. Application service providers offer services such as search, social networking support, and various other capabilities that can attract the consumers to their sites. Anything that an application service provider can offer can be offered within their service network by a mobile network operator. The decision to offer these services is based on the business case the operator may be able to make.

A somewhat controversial data monetization service is the generation of advertisements by the mobile network operator. The operator may have the ability to generate advertisements and show them on the mobile device of its subscribers. It may also be able to push advertisements either as text message, pop-up messages, or to a customer application running on the device. Sending advertisements to subscribers does generate revenue, but may also have the adverse impact of annoying the subscribers who may switch over to a carrier who generates none or a reduced number of advertisements.

There are many other innovative services that application service providers will create for consumers over the years. All of those are possible services that a mobile network operator can offer to its consumer subscribers.

CHAPTER 8

ENTERPRISE-ORIENTED DATA MONETIZATION SERVICES

In this chapter, we look at the data monetization services that a mobile network operator can offer to the other set of its users, namely, the enterprises. In the ecosystem description as discussed in Chapter 2, an enterprise is an organization that has several individual members, and these individuals all access the data services through the mobile network operator. Such an enterprise could be a business. For example, a bank or a car dealership would be an enterprise. They could also be local government (e.g., city employees or police department) or a national government agency or a not-for-profit organization. An enterprise usually has its own enterprise IT infrastructure, consisting primarily of data centers and services that are intended for its employees. The enterprise may negotiate group agreements for its employees with the mobile network operator to offer special plans or deals for them.

We begin this chapter with an overview of the model for enterprise access and how the mobile network operator may offer services to an enterprise. We also examine how a competing approach to providing the same services may be offered by a competing application service provider. Then we discuss the differentiators that a mobile network operator may have compared to the application service provider, followed by an overview of some services that can be provided to the enterprise.

Techniques for Surviving the Mobile Data Explosion, First Edition.
Dinesh Chandra Verma and Paridhi Verma.
© 2014 The Institute of Electrical and Electronics Engineers, Inc. Published 2014 by John Wiley & Sons, Inc.

8.1   MODEL FOR MOBILE NETWORK OPERATOR SERVICES TO THE ENTERPRISE

For the discussion about data monetization services that a mobile network operator can provide to its enterprise customers, we assume that the employees of the enterprise access the services within the enterprise using the abstract network model shown in Figure 8.1.

The left-most side of the figure shows the mobile device that is used and operated by an employee of the enterprise. Let us consider data that is headed from this device to a server located in the enterprise. This mobile device sends the data over the air to equipment located at the cell tower. From that point, the cellular network infrastructure transfers the data over to the service network of the mobile network operator. At the service network, the data packets may be processed by intermediary functions provided by the operator and are then sent to the public Internet. From the public Internet, the packets enter the enterprise network. From the enterprise network, the packets can enter a data center that is hosting the destination server. On the reverse path, the packets from the server reach the mobile data in the reverse direction.

The portion of the network from the cell tower till the service network is under the control of the mobile network operator. The portion of the network from the enterprise data center to the enterprise network is under the control of the enterprise.

Although Figure 8.1 depicts the path as a linear topology, the structure of the network is not completely linear. In most networks, there is a tree-like structure imposed on the cellular network side—with about an order of a few hundred thousand cell towers that follow more or less a tree-type structure into a few devices into the service network. The service network can be viewed as having a linear topology, although for bigger operators, the service network itself can be a fairly complex Internet Protocol (IP) network. The Internet itself consists of numerous independent clusters of large networks. The mobile network operator would be connected to the Internet at many

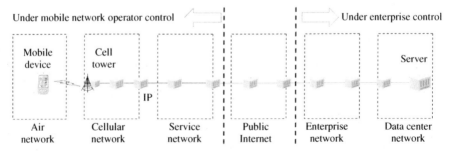

FIGURE 8.1   Enterprise access network model.

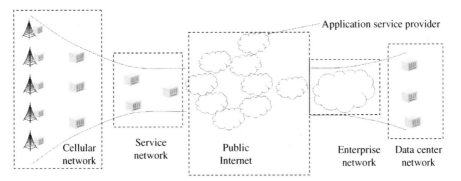

FIGURE 8.2   Structure of network for mobile access.

points, although the path from a single mobile device will likely go through the same interconnection point regardless of how the device moves in a session. The enterprise may similarly be connected to the Internet at many different points and may have more than one data center located within its network. This distribution and the structure of the network are shown in Figure 8.2.

The services that a mobile network operator provides to the enterprise can be offered from the service network, or they can be offered from locations in the cellular network itself. Note that the services to the enterprise may be provided by an application service provider or a cloud service provider that is based on the Internet. In that case, the location of that service could be in the position shown in Figure 8.2. While the packets flowing from a mobile device to the server in the data center would typically flow through the location of service provided by the mobile network operator, they will not necessarily flow through the location of the application service provider unless specific steps are undertaken to ensure that flow.

8.2   MOBILE NETWORK OPERATOR DIFFERENTIATORS FOR ENTERPRISE SERVICES

A differentiator is a property that the mobile network operator can provide to its consumers that others providing the same service do not have. As services are offered by the mobile network operator to enterprises, these properties would tend to promote the enterprise to get the service from the mobile network operator, rather than some other provider.

In order to identify the differentiators a mobile network operator may have in offering enterprise services, it is useful to understand the typical round trip latencies involved in communication between the mobile device and different parts of the network. The typical latency (round trip) on the air is less than 10 ms, and the latency involved in processing various functions at the cell-tower equipment would also be less than 10 ms. Communication between the cell-tower equipment and the Radio Network Controller may add another 10–20 ms round-trip latency depending on network configuration. The total latency of the 3G cellular network in accessing the service layer network would be somewhere between 50 ms and 100 ms, depending on network configuration [41]. The latency in the Internet could range between 10 ms and 100 ms, depending on the distance traversed. The latency within the enterprise network would typically be 10–20 ms, depending on the size and complexity of the network.

A mobile network operator can offer a service on behalf of the enterprise from within a location in its service network, or from a location within the cellular network, for example, from the cell tower or from somewhere deeper in the network. In offering any service, it has to compete with an application service provider that can offer the same service to the enterprise from within a location in the Internet. Because the mobile network operator is closer to the mobile device, it may have a latency advantage in offering some services to the enterprise.

Figure 8.3 shows a chart with the typical latency for a service offered from different locations to a mobile device. The figure assumes that the round-trip latency contribution from the mobile device is about 10 ms, and the latency of the service itself is about 5 ms. If the mobile network operator can deliver a service from the cell tower itself, then it has a significant latency advantage over comparable service offered by a competing application service provider. The cellular system is designed to make it enable the mobile network operator to deliver services from the service network. However, the latency advantage of the mobile network operator in delivering these services is not very significant compared to those of application

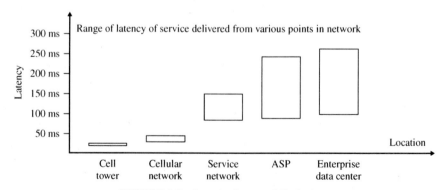

FIGURE 8.3   Latency from mobile device.

service provider. Only when the application service provider is located at a significant distance away on the Internet would there be some substantial differences. Similarly, the latency advantage as compared to the enterprise service itself is not a significant one.

If there are services that the mobile network operator can offer from within the cellular network infrastructure, ideally from the cell tower itself, it would have a significant advantage over any competing offering made from an application service provider.

Another differentiator that a mobile network operator has compared to an application service provider is in its ability to intercept packets from the mobile device going to the enterprise network. Since all of this data needs to be traversing the mobile network, the operator can intercept the packets and provide intermediary services. These services can be beneficial to the employees of the enterprise. On the other hand, the packets flowing to the enterprise will not naturally be directed to the service hosted by the application service provider. That type of redirection can be arranged, but it would typically require some carefully orchestrated configuration of systems in the enterprise as well as the applications on the mobile device.

Another advantage that the mobile network operator has over the application service provider is its ability to locate the employees of the enterprise without requiring any application to be installed in the mobile device. An application service provider would need some software on the mobile device to report its current position. Not having to manage the software for location determination in the mobile devices could be a significant simplification in rolling out any location-based service for the enterprise. Of course, some of the complexity for tracking the location is now shifted to the mobile network operator.

A mobile operator may be in a position where it enters into an exclusive arrangement with an enterprise to provide all of its employees with mobile services. If all the employees of the enterprise are going to get their mobile phones from the operator, all of the enterprise traffic is flowing through the operator's network, then it can provide various services in a transparent manner. However, if the relationship is not exclusive, and some employees may access the enterprise sites through a competing mobile network, the operator is basically equivalent to an application service provider—and needs to implement the same mechanisms for content redirection that an application service provider would. Differentiators like lower latency and transparent service interposition are features that will only be available to employees accessing enterprise services through the mobile network operator.

With these differentiators in mind, let us explore some of the services that a mobile network operator can provide to enterprises with a differentiated value proposition.

8.3 CACHING AND CONTENT DISTRIBUTION

The mobile network operator has two differentiating propositions that allow it to implement caching for any employees accessing servers on the enterprise network. The mobile network operator has the ability to transparently intercept packets flowing from the users to the enterprise server, and the employee's mobile device would typically be at a smaller latency from the caching location than going the full distance to the enterprise servers. In practice, many network operators implement transparent caching wherever they can, primarily to save the bandwidth flowing outside their network that provides them financial benefits. The capability exists in almost every mobile network operator. The key question would be how they can offer the caching services in a manner so that an enterprise will be interested in paying the mobile network operator for the same.

Caching helps the mobile network operator by reducing the bandwidth used on its network, but it also provides benefits to the enterprise by reducing the number of requests that are being handled by the enterprise server. If a smaller-capacity server can be used by means of the caching services provided by the operator, then the enterprise may be willing to pass along some of the savings to the mobile network operator. The caching function could be turned on dynamically only when desired by the enterprise to meet larger traffic volumes that they have planned the system capacity for.

The concept of improving scalability on demand by serving content from a different site is used widely by sites running on the Internet and is the basic value proposition of multibillion dollar content distribution networking (CDN) industry. Currently, CDN services are offered by application service providers running in the Internet, and sophisticated tricks with domain name systems and content rewriting are used to direct requests to a CDN site instead of the original system.

A mobile network operator can provide a service that offers all of the benefits of a CDN without requiring the mechanisms to direct traffic to its site. Since all requests from the subscribers of the mobile network need to traverse its network, it can intercept the requests and cache them when indicated to do so from the enterprise server. The approach does not work for users that are accessing the site from a network different than that of the mobile network operator. However, if a significant fraction of enterprise users access the network from the operator under some business agreement, the reduction in traffic load at the enterprise server can be achieved effectively.

One unique advantage that the mobile operator has compared to Internet-based application service providers is its ability to move content very close to the mobile users. If it is able to host content at the cell tower itself, the content is being located a few millisecond distance away from the mobile user. This can improve the user quality of experience significantly for the user for most accesses. Specifically, if the content happens to be streaming video or multi-media, the user experience of the video would be much better if it comes from a site with a small latency.

8.4 MOBILE TRANSFORMATION

In many enterprises, employees are provided access to their information services and systems from standard laptops or desktops. Since the rise of the browser in mid-1990s, they have been the de facto interface for employees to access enterprise services. Since the users had devices like laptops and desktops with sizable display screens, the websites and enterprise services were designed to use the large screen area optimally.

With the advent of Smartphones, the available screen size is much smaller. Even with an increased focus on larger phones, there is a physical limit of what size screens people can carry around comfortably on their person. Until the advent and adoption of some technology like projection of screens on user glasses, mobile devices are likely to have smaller screen sizes. In this case, many of the websites and services developed in the enterprise become cumbersome when displayed on the smaller screen. As an example, the navigation bar present on the left side of many websites becomes a nuisance when viewed on a Smartphone screen.

One solution is to create dedicated applications for the mobile phone and devices to enable access to enterprise services. While this is a very good solution, there is a significant effort involved in migrating all the services available on the web to their corresponding applications on the mobile phone. Furthermore, since desktops and laptops will continue to be used for a substantial period of time in the offices, the existing web-based systems also need to be supported.

The more pragmatic solution is to create an intermediary that detects if a user accessing a web-based service is coming from a mobile device. If so, the content of the pages are rendered in a manner so that it is more convenient for a mobile user. Otherwise, the usual web interface is shown. The intermediary acts like a transparent proxy for web-based employee access and acts like a content transformation engine for users on a mobile device.

A mobile network operator can offer to perform these types of mobile transformations for the enterprise users. Since all of the accesses from the mobile network users are from mobile devices, the transformation function can be always turned on, and the challenges involved in deciding whether a user is mobile are avoided. The mobile network operator can charge a fee to the enterprise operator for offloading this complexity to them. Since the mobile network operator will be providing the transformation capabilities to multiple enterprises, it can amortize the cost of developing the transformation system and provide a lower solution to the enterprises than if they were to do it themselves.

8.5 FOG COMPUTING

Within enterprise data centers and IT infrastructure, there are two technologies that have become widely adopted. The first is the use of virtualization technology and the other is the use of cloud computing.

An enterprise data center consists of many servers, which are used to run different pieces of the software. The structure of software running in a data center can be represented in a layered structure as shown in Figure 8.4. The physical machine hardware runs an operating system, such as UNIX and its variations. The operating system supports a middleware layer. Examples of middleware include database system, web application development system, message passing systems. The final enterprise software is implemented on top of the middleware.

In virtualization, data center software is run on virtual machines instead of directly on the physical system. Normally, a single physical machine runs a single operating system instance on which the data center software applications are run. In the case of virtualization, a software element called the hypervisor allows the creation of multiple independent operating system instances on the same physical hardware. In essence, a single physical machine acts like instances of multiple machines. Virtualization has many benefits, including the ability to dynamically reduce the power consumption in data centers, the ability to move computational elements around easily, and to reduce the number of servers required in the system overall (Figure 8.5).

Another technology that has been seeing increasing adoption is the concept of cloud computing. In the case of cloud computing, the servers are not

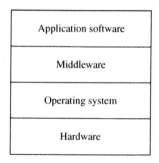

FIGURE 8.4   Structure of server software.

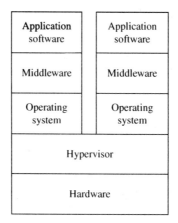

Application software	Application software
Middleware	Middleware
Operating system	Operating system
Hypervisor	
Hardware	

FIGURE 8.5   Server software with hypervisors.

hosted as physical machines within the enterprise itself, but rented from an outside application service provider as needed. For computation that may not always be needed, for example, if the enterprise only needs to run some type of analysis software once in a couple of months, it may be more cost-efficient for the enterprise to rent the computational capability from a cloud service provider instead of trying to have servers of its own.

Cloud computing from application service providers is usually offered in three flavors—infrastructure as a service (IaaS), platform as a service (PaaS), or software as a service (SaaS). In IaaS model of cloud computing, the enterprise can rent virtual machines on the cloud. In PaaS, the enterprise rents middleware infrastructure from the cloud, and in SaaS, the entire application is provided from the cloud.

A variation of cloud computing, called fog computing [42], has been proposed by some researchers to address issues associated with performance and scalability. In cloud computing, the enterprise applications run from a shared infrastructure on the Internet somewhere. In fog computing, the concepts of cloud computing are offered from a system that is much closer to the end-client. The term fog computing is used because it is a cloud that is much closer to the ground, resulting in fog. Another variation on the same idea in literature [43] uses the term cloudlets.

A mobile network operator can offer fog computing services to the enterprise customers in either the IaaS, PaaS, or SaaS model. In the IaaS model, the enterprise rents virtual machines from the mobile network operator. These virtual machines can be provided at the service network of the operator or at the cell tower itself. A virtual machine provided at the cell tower is much closer to the end-users than the enterprise data center can get in other ways and is a differentiating proposition that only the mobile network operator can provide. Similarly, in PaaS, the mobile network operator can provide a fixed set of middleware options, while in SaaS, a set of complete applications are provided.

Distributing the cloud to individual cell towers does have a complexity associated with it, which can reduce the potential benefits of the latency and performance gains of fog computing. Managing distributed applications is more complex than managing applications that are located in a centralized cloud and require a higher level of automation. The cost benefits of a large cloud operating at scale may be reduced by distributing to the cell tower. For various applications, the tradeoff in cost and performance needs to be taken into account to determine if fog computing is appropriate for an application in a given network environment.

In essence, the concept of fog computing is to offer cloud computing services from a location in the mobile network operator's environment. By making it closer to the end-user, fog computing offered by a mobile network operator can deliver all the benefits of cloud computing to the enterprise, with the added benefit to improve user response time and quality of experience.

8.6   LOCATION-BASED SERVICES

Location-based services were discussed as a possible option to offer to consumers by a mobile network operator. The differentiators in offering location-based services that a mobile network operator has are the same for consumers and enterprises. However, when the mobile network operator has a relationship or agreement with an enterprise, it can offer specific location-based services to the employees of that enterprise. It can also offer location-based services that improve the operation of an enterprise. Some examples of location-based services that a mobile network operator can provide are as follows:

- *Tracking of assets*: Many enterprises have different types of assets that may be distributed over a wide area. As an example, a cable company has its repair crew and trucks moving around in an area. A sanitation or waste management company may have dumpsters that may be distributed in various portions of a town. Some of these dumpsters may be rented out to specific residences or businesses on a short-term basis. A forestry company may have its wood chippers and tree cutters distributed across a wide area of forest. All of these enterprises need to know the current location of their assets and where they are situated. A mobile network operator can offer the service of tracking such assets for them. Assets can be tracked by having a module with a GPS unit or a SIM card attached to the asset, and the phone company can record the location of the unit.

- *Geofencing*: One specific instance of asset tracking is geofencing. Geofencing tracks the location of an asset. As long as the asset is within a specified area on the map, the system does nothing. However, if the asset leaves the boundaries of the specified area, an alert is provided to the enterprise. Geofencing is useful in situations where the only interesting case for tracking is if the asset moved outside the designated area. In addition to tracking, if assets have moved outside a fence, one can also track if assets or people have moved inside a fence, for example, identify if an employee has arrived at a work-location. Geofencing has applications beyond the enterprise, for example, it can be used to track pets for owners if they wander away from a locality, tracking wild animals or herds of cattle, or tracking the location of unaccompanied minors in schools.

- *Fleet management*: For many enterprises, it is important to operate and manage a fleet of vehicles. Examples of such enterprises include taxi companies, pickup and delivery companies, and bus/train operators. For most of these enterprises, tracking where the members of their fleet are, and deciding which fleet member ought to be routed to a new request, is important. As an example, consider a taxi company where people request for a pickup and drop-off at random intervals. When a new request comes in, a decision needs to be taken as to which of the available taxis ought to be dispatched. Knowing the location of the fleet at all times and making decisions on dispatching them to new requests for pick up are important services for them.

- *Emergency management*: When an emergency such as a snowstorm, flood, blizzard, or earthquake happens in an area, most enterprises take charge of the task of tracking their employees and ensuring that they are in a good state. This requires tracking their location, informing them using a call or a text message, and asking them to report back to the enterprise about their situation. Since normal enterprise access, for example, electricity or Internet access from home may not work in this situation, the mobile network provides a convenient alternative option provided its cell sites remain operational during the emergency.

These are just a few of the examples of location-based services that a mobile network operator can offer to enterprises. More such services can be envisioned and added to the list. Regardless of the specific service that one is considering, there are two alternatives for any enterprise to avail of such services. The enterprise can operate the service on its own premises, or the enterprise can subscribe to such a service offered by the mobile network operator.

When the service is operated by an enterprise on its own, it is just using the mobile network operator to transport its bytes over the network. In some cases, the enterprise may get the location of its employees or assets from the network operator, in other cases, it may get location from its own infrastructure, for example, applications on the mobile devices of employees, or its own tracking modules on the assets. In this case, the network operator will have less of a data monetization opportunity. In order to monetize the service, the network operator needs to offer this service in a way that the enterprise will find it better to obtain it from the network provider rather than running its own or getting the service from an over-the-top service provider.

For small-scale enterprises, which do not have the resources to develop their own set of services, the mobile network operator may be able to offer these services more effectively. Similarly, if the mobile network operator is able to offer these services, for example, tracking people and fleets without explicitly requiring application on people's mobile devices, many enterprises may find it attractive to use the services from the operator.

8.7   SECURE HYPERVISOR SERVICES

One of the challenges that many enterprises face today is that of enterprise data and application security when their employees are using their own mobile devices or using the enterprise-issued mobile device for personal uses. Mobile devices, while very convenient, have the challenge that they are easy to misplace and expensive models are likely to attract thieves. Furthermore, when an employee is installing and putting applications on the mobile device that are not completely within the control of the enterprise, leakage or loss of sensitive enterprise data due to presence of malicious applications on the device can happen.

One solution to this problem is to have two personas on a mobile device using a mobile device hypervisor which can allow two virtual devices on the same physical device, one of them for enterprise usage and the other for personal usage. In this case, the mobile network operator can offer an additional service to the enterprise in that only the virtual device that is running enterprise applications is allowed to connect to the enterprise network. This approach secures access to enterprise network, but does not prevent loss of enterprise data due to a misplaced device.

Another solution to this problem is an application of fog computing. The enterprise applications never run on the physical device, but are instead executed on a service in the network, and only the visual rendering of the applications are shown on the screen of the device. With this approach, the enterprise applications and data never leave the network, and are protected even if the device is lost.

In order to get good performance, the hosting service needs to be close to the mobile device. The mobile network operator, if offering this service from the cell tower, is in a position to deliver the best response time. It can also be hosted in the service network, although then the only advantage of the network operator over an Internet-based application service provider is that it can intercept the flows transparently for enterprise employees using its network.

CHAPTER 9

APPLICATION SERVICE PROVIDER-ORIENTED DATA MONETIZATION SERVICES

In this chapter, we look at the data monetization services that a mobile network operator can offer to the other set of its users, namely, the application service providers. An application service provider in some instances may be competing with the mobile network operators for the services and, in some instances, may be a customer of the network operator. Therefore, it is useful to understand the model for the offering of the services by these two different constituencies in the mobile data ecosystem and the differentiators that the mobile network operator can bring to the application service provider. The discussion of those issues is the first section of this chapter.

The rest of the chapter looks at different services that the mobile network operator can provide to the application service provider. Many of these services are variations of the service that the mobile network operator can offer to an enterprise. However, the benefits of a service as perceived from the application service provider perspective may be very different than that provided from the enterprise perspective.

Techniques for Surviving the Mobile Data Explosion, First Edition.
Dinesh Chandra Verma and Paridhi Verma.
© 2014 The Institute of Electrical and Electronics Engineers, Inc. Published 2014 by John Wiley & Sons, Inc.

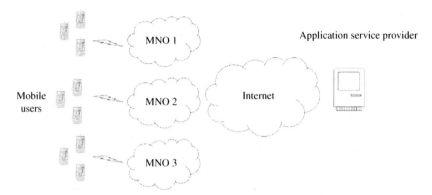

FIGURE 9.1   Model for access to application service provider.

9.1 MOBILE NETWORK OPERATOR DIFFERENTIATORS FOR APPLICATION SERVICE PROVIDERS

The manner in which applications are implemented on the mobile network is very similar to the model of applications for the enterprise shown in Figure 8.1. However, there is one important difference between the two. In the case of the enterprise, the mobile network operator is able to negotiate a situation where all mobile users (the employees of the enterprise) would be accessing services through its network. Even if the relationship is not exclusive, the mobile network operator can offer its services as that provided by an application service provider for enterprise users coming from another mobile network.

However, the application service provider usually would have little or no control over the choice of the mobile network providers for their users. Any mobile network operator is only able to service customers along the path of a subset of the users of an application service provider. In any geographical region, there are usually two or three dominant mobile networks. Figure 9.1 shows such a situation, where there are three mobile networks in use. If all of the nine Smartphone users are accessing a site operated by an application service provider in the Internet, each of the three mobile networks are only able to intercept traffic belonging to three of those users.

A substantial fraction of the users of any application service provider will not be using the mobile network provided by any individual operator. In order for a service provided by a mobile network operator to be attractive to an application service provider, the service needs to be valuable even if only a fraction of application service provider's users avail it. Some services that a mobile network operator can offer to enterprises will not be as attractive to application service providers because of this reason, mobile content transformation being such an example.

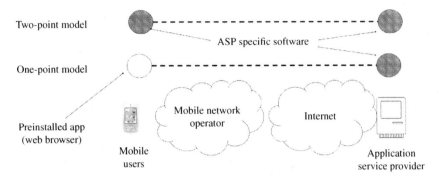

FIGURE 9.2   One-point and two-point models for applications.

Before discussing the differentiators that a mobile network operator has, let us take a quick look at the way in which application service providers provide services to mobile users. There are two models that an application service provider can use. We refer to them as a single-point model or a two-point model. Both these models are shown in Figure 9.2.

In a single-point model, the application service provider would run a service that is accessible by a preinstalled application on the mobile device. The application service provider does not own or create the preinstalled application. In most cases, that preinstalled application is the web browser on the mobile device, and the application service provider would simply operate a website to provide a valuable function to their customers. In the two-point model, the application service provider would have an application on the mobile phone in addition to the website or service it operates on the Internet.

Using the one-point model and just providing a service or website reduces the development costs for the application service provider. However, the two-point model having a customized application on the mobile device allows development of applications with better user experience and enables new capabilities that may not always be possible with a preinstalled application such as the web browser. In the two-point model, the communication protocol used between the application and the service can also be customized to the needs of the application and need not be constrained with the limitations of the standard protocol.

In such an environment, there are only a limited set of differentiators that a mobile network provider can offer to the application service provider. The first such differentiator is the ability of the mobile network provider to have a presence at a site with a very low latency to the end-user. As discussed in the preceding chapter, a mobile network operator has the ability to host services and content right at the cell tower at a latency that is an order of magnitude lower than that of going across to the Internet. This reduced latency to the end-user would be a key attribute that application service providers cannot get any other way and could be exploited by the mobile network operators.

Another differentiator that many mobile network operators have is their size and the number of subscribers they support. While this may not be a competitive differentiator against large application service providers, this scale could be quite attractive to many smaller application service providers. Let us consider the case for United States, where there are a few large mobile network operators. The United States also have a few very large application service providers, for example, popular social networking sites and web search engine providers. The larger application service providers have a user population that is comparable and, in some cases, larger than the subscriber population of any mobile network operators. However, there are many smaller application service providers that do not have the same number of subscribers. If a large mobile network operator were to offer some service based on aggregating information from their subscribers, it may not appeal to the group of large application service providers who may be able to collect equivalent or better information on their own. However, for smaller application service providers, this service may be very attractive.

Consider the example of a start-up that has decided to offer some type of enhanced routing and mapping service, and thus in competition with one of the large application service providers, which also provides a similar service. The start-up needs to obtain information about current traffic congestion on various roads. A large mobile network operator may be offering such a service by aggregating information from the location of its subscribers. Similarly, the large application service provider can offer this service by aggregating information from mobile phone users running an application provided by the larger application service provider. The start-up is competing with one of the services provided of the large application service provider and may prefer to obtain these services from the mobile network operator instead.

With these differentiators in mind, let us explore some of the services that a mobile network operator can provide to application service providers with a differentiated value proposition.

9.2   CACHING AND CONTENT DISTRIBUTION

As in the case for caching and Content Distribution Network (CDN) discussed in Chapter 8, the mobile network operator has two main advantages when offering caching and CDN services for application service providers: it has the ability to transparently intercept packets flowing from the users to the application service provider server and it can locate its caching at a smaller latency from the customers than going the full distance to the application service provider.

As in the case of enterprise systems, caching can reduce the number of requests that are being handled by the application service provider. This can reduce the capacity needed by the application service provider, and the capacity reduction could be substantial even if only the requests from the customer set coming through the mobile network operator can be handled. For application service providers that send out large volumes of video or audio, which are all cacheable content using up a large amount of bandwidth, the reduction in capacity needed at the core site could be substantial.

9.3   FOG COMPUTING

Fog computing is, broadly speaking, a generalization of caching/CDN principles to general applications. Like the former, it is an example of a service that can deliver a significant value to an application service provider even if it is provided and available to only a fraction of its customers. In this model, the service that the application service provider would have been running in their data center would instead be running inside the infrastructure of the mobile network operator. The approach would save both processing capacity on the server of the application service provider, as well as the bandwidth traversing the various networks. Offering fog computing as a service to data-intensive application service providers could be a win-win situation for mobile network operators.

As an example, let us consider an application service provider that offers streaming video over the Internet. If video is streamed from the servers on the Internet, it is coming from a site that is further off from the network, resulting in increased stalls and a lower throughput than if it were coming from a server hosted inside the mobile network infrastructure. The user quality of experience would be improved substantially while the load on both the network and the application service provider will be reduced. This would work even if only a fraction of users of the application service provider used the mobile network operator's access.

Figure 9.3 shows the scenario of video streaming hosted as an instance of fog computing provided by one of two mobile network operators. MNO 1 provides fog computing services, so video download for its customers comes from a site with a much lower delay. The video streaming service is faster and better performing for the users of this mobile network operator compared to those who do not offer this service. This may provide an incentive for users to prefer the mobile network operator that is offering the fog computing services. The application service provider and mobile network operator both benefit from the use of fog computing model.

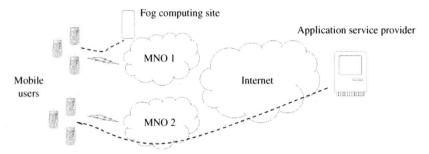

FIGURE 9.3 Fog computing for application service provider.

There is another nontechnical advantage in an application service provider leveraging the fog computing services offered by a mobile network operator. In many countries, mobile data is paid for on a usage basis, and unlimited data plans are disappearing. As mobile device usage grows, the data charges associated with watching a movie or video on the Internet can act as a damper on the use of data-intensive mobile applications such as streaming video. By offering a monetary discount to users who are accessing the services from the mobile network operator services, the demand for an application service provider can be boosted up. The increase in revenues that results, that is, the economic benefit, can be used to pay for the hosting of the infrastructure, ending up in a win-win situation for both the mobile network operator and the application service provider.

The same approach applies to other services that application service providers offer. In general, the benefits of moving computation from the data server to a location within the mobile network operator will be better under the following set of conditions:

- If there is a significant amount of data being transferred by an application.
- If there is a significant impact of network latency on user performance.
- If the application is chatty, that is, it makes a number of short sequences of request response interactions with the server.

As cloud computing is offered in different flavors, fog computing offered by the mobile network operator can also be offered to application service providers in different flavors. In the context of infrastructure as a service, fog computing consists of a mobile network operator offering virtual machines for use by application service providers within its infrastructure. This reduces the latency of the hosted applications. At the same time, it provides a well-defined service that the mobile network operator can offer.

A similar model can be offered by providing a platform as a service by the operator. This could be useful if any platform for developing applications became common across different application service providers. However, since applications are the unique attribute among different application service providers, hosting specific applications as a common service for different application service providers may not be an effective method for data monetization.

9.4 INFORMATION AGGREGATION

One of the crucial pieces of information available to the mobile network operator is the location, change of location, and some demographical information about each of the mobile network subscriber on their network. While the specific information about any single individual is sensitive and may be shared only under strict privacy regulations, the information when aggregated can be quite valuable and provide useful resalable information that will not be subject to the same strict regulation.

The most obvious information that can be aggregated is a mapping of the location of users and how fast that location is changing along different segments of a roadway. A mobile network operator can track the speed at which its subscribers are moving along the highway. This information can then be used to provide traffic information on the roadways to companies that operate navigation devices and systems.

Comparable traffic information can be collected by very large application service providers who have their own maps and navigation service. However, this does require the wide prevalence of Smartphones and the willingness of users to enable battery-hungry location tracking applications. There are many application service providers that could benefit from this information, but do not have the broad penetration of the bigger application service providers to get this information. They could be potential market for this type of information aggregation.

An advantage that the mobile network operators have is that they can track and aggregate the location information from the cellular triangulation of different users on their network without requiring any application on the mobile device. This saves the battery on the Smartphone as well as allows the tracking to be performed on users of feature-phones. In localities where Smartphones are rare, for example, Africa, the mobile network operator is the only one who can obtain this type of aggregated information.

The mobile network operator also has the option of using a coarser but easier way to determine the location of users. They can simply track the cell tower to which a user is associated with. The accuracy of the location would depend on the coverage area of each cell; it will be much finer resolution with picocells and microcells than with macrocells. As people switch calls across cell towers, new call data records are generated. By tracking the trajectory of the mobile devices through different cell towers, associating them with their rate of movement and overlaying the information on the network of roads in the area can give a reasonably good estimate of the traffic flowing on different roads in any area.

The traffic information is not the only type of aggregated information that the mobile network operator can provide. In many cases, the mobile network operator has information about the demography of a caller, for example, their age, their place, or area of residence. In some cases, for example, when they perform credit check on their users to provide the service, they would also have access to their credit scores or income information. When people are availing of discounted rates for phone-plans, for example, using an employee or organization group rate, the network operator knows about the user organizational affiliation. The mobile network operator can determine the demographic distribution of different characteristics of the population. Aggregation of user information eliminates many of the privacy concerns that people have about individual information. Demographic distributions can be valuable for many application service providers, for example, advertisers that rely on location-specific advertising, for example, placing advertisements on billboards.

Information aggregation can also identify the commute patterns existing in a community and detect the current location of trains or buses by correlating the movement of users with the train or bus schedule and their past patterns of travel. This can provide a service for tracking buses and trains, a capability that is not developed well in many emerging markets.

There are several types of application service providers who will be happy to obtain the current or historical pattern of people movements across a specific geographical area. This will allow them to understand how to customize and offer their services to their customers.

9.5   INFORMATION AUGMENTATION

Another useful benefit that a mobile network operator can provide to the application service provider is that of information augmentation. As an example, many application service provider websites offer location-specific advertisements on their site. If you are visiting the website, the application service provider collects your machine's Internet Protocol (IP) address and tries to relate the IP address to a rough geographical location using a variety of techniques [44,45], but mostly a database of lookup tables mapping addresses to locations. Although approximate and not very accurate, the information is still granular enough to identify a computer within 20–30 miles from its actual location.

When the same techniques are used to identify a user's location when the user is coming from a mobile device using Long Term Evolution or 3rd Generation Partnership Project protocols, the IP address normally is associated with the Internet address pool allocated to the mobile network provider. In most cases, they can only be mapped to a few discrete points within the country, or in some cases to a slightly finer granularity. This seriously hampers the ability of the application service provider to identify the location of a user.

Application service providers who have their native applications on the mobile devices would be able to track the location of the user. However, many application service providers will be accessed by users without downloading their applications. If permitted by the applicable regulatory guidelines, the mobile network operator may include the approximate location of the user in the messages being passed to the application service provider. The mobile network operator may be able to charge a fee for providing this extra information to the application service provider.

In addition to the location, the mobile network operator may be able to pass other information, for example, the native language of the user. This would allow the application service provider to choose the language in which its content is displayed preferentially. The primary restriction on such information augmentation would be any applicable local regulations, as well as any perceived adverse reaction from the customers of the network of their location identification.

Another type of information that the mobile network operator can pass to the application service provider is the type of network connectivity a user has to the network. As an example, if the radio spectrum is constrained in the environment, the mobile network operator can pass this information to the application service provider. The application service provider can adapt the type of information provided to the users, depending on the network conditions of the user. As an example, consider an application service provider operating a website that is graphics heavy. If the user has poor network connectivity, the application service provider can give a version where the pictures are of lower resolution. It enables the application service provider to better manage the user experience of anyone accessing their websites depending on their current network conditions.

9.6   HISTORICAL INFORMATION-BASED PLANNING

Location-based services were pointed out as a good source for creating data monetization services oriented towards the consumers as well as enterprises. For application service providers, and for some enterprises, historical location-based services can also provide a useful source of revenue.

The mobile network operator has access to the history of a large portion of the population and how it moves about. Mining that information can yield a wealth of interesting insights that can provide valuable information to other application service providers. In this section, we look at some of the typical scenarios where such cases can be valuable.

- *Transportation Planning*: When public transportation in a metropolitan area is planned, it is usually based on rough guesses made around anticipated population centers and their tendency to travel towards likely destinations. However, the mobile device has provided a way for the mobile network operator to collect and store historical patterns of movement of people around an area. By tracking the position and trajectory of the subscribers, the mobile network operator knows exactly what the travel patterns of the people are, both during their commutes back and forth from the work, as well as the travel patterns during off-peak hours. When new buses or train lines need to be scheduled, this information can provide the real snapshot of the demand for these buses and trains. This can help plan for new bus-routes, create new schedules for buses and trains, and even determine if a new extension of a train line is needed.

- *Facility Location*: Knowing the actual travel patterns of people can be a valuable insight into deciding the right venue for locating a new facility, where the facility may be a new shop being opened. For each new shop location, it is advantageous to place it near areas of high traffic, preferably among areas with high traffic of a desired demography of users. Historical location data can be useful for deciding where the next coffee shop ought to be opened.

- *Marketing and Advertising*: In any city, there are multiple billboard locations. However, it is not clear how many people are crossing by the location of any particular billboard. The location of the billboards is normally based on a crude guess of where people may be expected to pass by. However, with the cell phones, a mobile network operator can keep an accurate count of who has passed by any given billboard. Having the actual historical number of traffic in any area can provide a very useful and valuable insight that can determine the value of each billboard as well as identify the best demography it ought to target.

- *Competitive Intelligence*: Many shops are large enough so that cellular location information can identify whether or not an individual is within the shop. This information can be used to provide competitive aggregated statistics about the number of users who are at that shop, the typical location where these users come from, and how much time they spend in the shop. The same information can be obtained about users who are at a competing shop. Such comparative analysis of traffic in the shops can be very valuable for many different shops, and the mobile network provider can monetize its ability to provide this information.

Innovative mobile network operators will probably dream of many more interesting data monetization applications that they can offer. These were just an enumeration of some likely use-cases of these applications.

SECTION 3

TECHNIQUES FOR ENTERPRISES AND APPLICATION DEVELOPERS

CHAPTER 10

AN INTRODUCTION TO MOBILE APPLICATIONS

In this and the next few chapters, we switch our perspective from a mobile network operator to that of an application service provider which is offering some services to be accessed through a mobile phone. In most cases, such an application service provider would need to have two components, a mobile application in the Smartphone and a server-side application software which communicates with the mobile application in the Smartphone. Mobile applications, especially mobile applications that exchange data with a server located in the Internet, are the primary source of the increase in mobile data.

The design and behavior of the mobile application can have a significant impact on the user experience of a consumer accessing the service, as well as on the use of resources such as Smartphone battery life, network bandwidth usage. As more and more application service providers attempt to make their services available to the users of Smartphones, the lack of adequate bandwidth in the network can be a significant issue affecting user experience. In this chapter, we look at the general topics involved in the development of mobile applications, discuss the difficulties in application development that are specific to mobile applications, and their impact on the application performance.

Techniques for Surviving the Mobile Data Explosion, First Edition.
Dinesh Chandra Verma and Paridhi Verma.
© 2014 The Institute of Electrical and Electronics Engineers, Inc. Published 2014 by John Wiley & Sons, Inc.

10.1   ANATOMY OF MOBILE APPLICATIONS

In general, the term mobile application refers to the piece of software that is resident on the mobile phone and interacts with the user of the phone. In some cases, this software is a self-contained component that does not interact with any other component on the network. As an example, applications like a flashlight or a virtual spirit-level on a Smartphone have no need to interact with any other system on the network. Apart from the instance where these applications are downloaded from the network for installation, these applications generate little load on the network. They may sometimes need to check for updates in versions, but largely function on a stand-alone basis.

Most applications on the mobile phone, however, need a fair bit of information exchange with one or more servers on the Internet. A news application needs to get the latest news from a website, puzzles need to be updated, games report and update high score values, multiplayer games need to synchronize game state across multiple players, address-books need to be synchronized, video and audio need to be downloaded, etc. It is in the exchange of this information that a paucity of bandwidth in the network can cause troublesome user quality issues.

If we look at the lifecycle of a mobile application, we can identify the following different stages as shown in Figure 10.1. In the initial phase, the application is developed. The development is followed by a phase of testing its performance and functions on the different platforms that it ought to run on. After the testing is done, the application is released. The release process consists of making the application available for download in one or more application stores. Once the application is made available to its users, and is

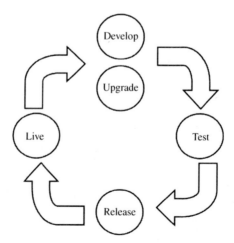

FIGURE 10.1   Life cycle for mobile applications.

live, it needs to be supported. The live phase of the application would typically be the largest one. After some period, an upgrade may be desired of the application in which case the upgrade restarts the cycle at the development phase.

Each of these phases may in turn consist of several different stages. From the perspective of software engineering, the development phase is most interesting and can be broken into various stages of application development in normal software engineering practice, for example, requirement analysis, design, development. The test phase may be the more time-consuming and expensive part of the process, given the large number of devices that need to be supported. The release process is generally streamlined and specific to different application stores. The live phase is when the issues associated with the limited resources in the network and the mobile phone come to the forefront, and is the phase where all the design choices made during development are put to the test.

Design choices made during the development phase impact all subsequent phases, and have a significant impact on the user experience and the capabilities of the application during its live operation stage.

10.2 TYPES OF MOBILE APPLICATIONS

A mobile application can be decomposed into three distinct components as shown in Figure 10.2. The mobile application consists of two pieces of software, a mobile component that runs on the mobile device and a server component that runs on the servers operated by the application service provider. These two components communicate by means of a selected protocol. Thus, the mobile application effectively consists of three components: the mobile component, the server component, and the intercommunication protocol. Based on the characteristics of these, three broad categories of mobile applications can be identified.

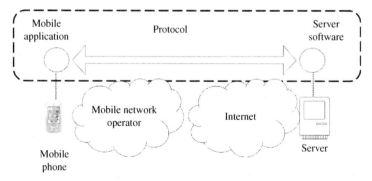

FIGURE 10.2 Mobile application structure.

The first category makes the decision that the mobile component consists of nothing but the web browser available on the mobile device and that the protocol used to communicate between the two components will be HTTP. This category is usually called a web application. The big advantage of having a web application is that it only requires development and support of the application on the server side. Since only one type of server will normally be supported (e.g., an organization may choose to run applications using a specific brand of web server running on a specific operating system on a bank of specific model of servers), it is more efficient and simpler than supporting several types of clients on which the mobile component has to run. The task of the mobile component is done by the browser, whose maintenance and upkeep is not the responsibility of the application service provider.

Note that the use of web-based application does not completely alleviate the problem of managing the diversity of different devices. Different devices have varying level of support for running web-based applications, which will need to be taken into account during the development and testing phases.

Web applications written on the server side using programs such as PHP or Python often provide client-side applications using a scripting language like JavaScript™ and use HTML, the markup language associated with documents exchanged using HTTP, to tie together the client and server-side applications The JavaScript programs are downloaded by the mobile phone's browser and executed locally. The most recent version of HTML is version 5 or HTML5, which mandates different types of capabilities on the browser including the support for streaming video and audio, the capability of local cached storage, and support for geolocation information. These capabilities allow a rich set of applications to be run and executed on the mobile phone using nothing but the standard browser on the client side. Like other content, the presentation look and feel of the application can be controlled by cascading style sheets or CSS.

If the mobile device needs to access any data or routine on the server side, these routines can be invoked as server-side scripts via HTML user controls (e.g., as forms or buttons) and developed in any language used for server-side scripting.

The second category of applications eschews the use of HTML for communication between the mobile component and the server-side component, opting for its own private protocol for communication and creating a mobile component that is independent of the browser. In this case, the application service provider needs to develop both the mobile component and the server-side component, as well as determine what protocol to use for communication between the two. These are referred to as native applications.

Native applications offer some advantages to the application developer over web applications. They can use more efficient protocols, for example, avoid the overhead of bulky HTTP headers or the need to encode all data being exchanged in HTML. They can also use the services on the native machine without being constrained by the restrictions imposed upon them by the web browser execution framework. By using the native language supported on the platform, applications can be more real-time and provide a much more rich and responsive visual interface than web-based applications.

The flip side of native applications is that they need to be developed for the platform, and the large number of devices available in the marketplace makes supporting them across several platforms very difficult.

In between these two models of building applications, there is a hybrid model which creates a native application but uses web-based protocol to communicate with the server-side component. This allows the hybrid model to choose the best points of both the native and the web application model.

Regardless of the model used for developing the mobile application, there are some challenges which need to be solved along the way of developing the application. In the next few sections, we look at these challenges and some of the ways to address those challenges.

10.3   DEVELOPING FOR MULTIPLE PLATFORMS

The first challenge in developing an application for the mobile devices is that there are so many of them. The marketplace balance among different platforms swings over time, and new mobile operating systems will continue to emerge. The Wikipedia page for mobile operating systems [46] lists more than 30 historical, current, and emerging mobile operating systems.

The plethora of operating systems and phones that are available makes it a challenge to develop a mobile application, and creates issues in the development, testing, and live phase of the system. Even if one were to focus on only one single operating system, there are multiple devices that support that operating system, and each device has its own form factor and set of resources. Applications need to be developed for each of the supported platforms, and also tested for proper functioning on those platforms. Managing the application on a large number of platforms is a challenge. It may work fine on one platform, but show a problem on another. The user experience may be adequate for one platform but not be very nice on another device with a different screen size.

The situation is not helped by the fact that each of these platforms has its own model and preferred programming language for developing applications. In some platforms, development tools are geared towards writing a program in variations of a programming language like C/C++, on other platforms the tools are geared towards writing a program in variations of another language. In essence, putting a native or hybrid application on each of the different operating systems is almost like developing several independent applications, which can add significantly to the cost and time required for development and testing.

The situation is alleviated somewhat when one is developing web applications, since the only requirement there is that the platform support a browser running JavaScript. However, there are some differences in JavaScript support between different platforms, which need to be taken into account. The JavaScript code needs to check for the browser version it is running underneath, and execute different sections of code in order to support the nuances of that particular browser version and platform.

In general, when there is a need to support multiple platforms, two approaches can be used to reduce the amount of effort required to support a single application on these platforms. The first approach is the use of an intermediary layer, and the second approach is the use of a translator.

In the use of the intermediary layer, a common application programmer interface (API) is defined. This API will be supported on each of the platform. A single application can be written on top of the common API. The common API is implemented by an intermediary layer, which will be different for each of the platform. Although the intermediary layer needs to be supported on each of the platforms, applications can be written in a single manner and supported across multiple platforms. If one has to create multiple applications, then the use of the intermediary layer can reduce the amount of effort required in development. This approach is illustrated in Figure 10.3.

When using the intermediary layer and a common API, it is important to insulate the application against the differences in the different platforms. This means that the applications need to be written, debugged, and developed using nothing but the common API. This approach also

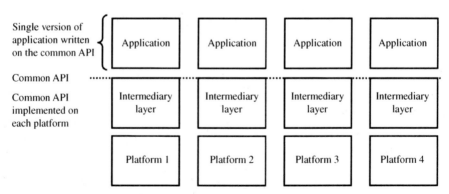

FIGURE 10.3   Intermediary layer approach for platform diversity.

precludes the use of any useful tools or software libraries that are available only on one platform. Creating the entire programming framework which supports just the common API and runs on top of many platforms could be a difficult task.

The other approach to manage multiplatform support is the use of a translator. In this approach, the application is developed on one platform, and then automated translation is done from this platform to all of the other platforms. As an example, an application can be developed using the web-development paradigm, that is, written in JavaScript with HTML and CSS. A translator would convert this code automatically into the native or hybrid application running on different platforms.

These two approaches alleviate the challenges associated with using multiple platforms during the development stage. There are complexities associated with testing phase for proper operation of the application on many different platforms. Many tests need to be run on each of the platform to check that the application is working properly. Test automation suites help to run several tests automatically. A careful design of the test suites can reduce the number of tests that need to be run. Another logistics challenge with testing for multiple platforms is the availability of the different platforms in-house. There are companies which provide virtual or physical appliances for testing purposes that help in obtaining the diverse set of devices that are needed.

During the live phase of an application, different customers using the application on different platforms could run into different problems. The help-desk staff for the application needs to be prepared for the various problems on each of the platform. In many cases, mobile application developers dispense with telephone help-desk because of associated costs, and only provide web-based help information. Even in these cases, the information about common issues on different platforms needs to be created and maintained. The lack of proper help and support infrastructure during the live phase of an application can cause serious detriment to customer satisfaction.

10.4   OPERATING SYSTEM VERSION MANAGEMENT

A related, but distinct problem from the issue of multiple platforms is that of managing an application through different versions of the operating system versions on mobile devices. New releases of operating systems can come in frequently; as an example between 2010 and 2012, over 10 releases of a popular operating system have come out [47]. An application developer who released an application at the beginning of 2010 had to decide the proper course of action to be taken every time a new version of the operating system is released.

In one sense, each release of the new operating system is like a new platform which needs to be supported by ensuring that the application works well on that platform. The functioning of the application needs to be tested on the new platform, and appropriate regression tests ought to be run. Any issues, either related to functions, or related to issues like performance or security need to be re-examined for each new release of the operating system.

Since not every device is likely to be upgraded to a new operating system even if it is available, most application developers need to decide which range of the particular operating systems they support. To maximize the appeal of their application, they need to support the latest version of the operating system, as well as a few back-level versions.

10.5   LIMITED RESOURCES

One of the biggest challenges when developing applications on a mobile phone is that the resources available on the mobile phone are limited. The computation power, memory and storage resources, as well as the battery power on the mobile phones are precious resources. While technological advances are being made in all of these areas all the time, the resources on the mobile phone are likely to be limited for a significant amount of time.

The solution to limited storage on the mobile phone is to augment it by additional storage on a server in the network, and the solution to limited computation power is to offload complex calculations to a more powerful server in the network. If the network bandwidth were plentiful, one could ignore the issues involved with the limitations of storage and computation. It is easy to build abstractions and infrastructure that could simply move things back and forth across the network as the need arises if there is abundant bandwidth.

When bandwidth is limited, or when bandwidth is associated with a usage charge, as in many of mobile phone data plans, the impact of storing items on the cache or the expense involved with offloading computation ought to be taken into consideration. An application that costs a lot in bandwidth charges may not appeal well to the end-users.

In order to manage the limited resources, several techniques and practices can be used. In the subsequent chapters of this book, we look at some of these techniques and practices.

10.6 GENERAL APPLICATION DEVELOPMENT CONSIDERATIONS

While the previous few sections considered issues that are unique for mobile applications, we must keep in mind that mobile applications are software applications after all. Therefore, all the general principles and best practices that are applicable to development of software in general are applicable to development of mobile applications.

Several of these best practices and principles are described in various books [48,49]. Most of these software practices are developed for large teams that are working on a big project. Mobile application teams tend to be much smaller. Nevertheless, several of these practices in software engineering are worth following even for the typically smaller team sizes.

Some of the practices worth following are maintaining software version control system, and following a set of well-defined processes for tracking requirements and changes. Software version control systems allow one to keep track of changes, and identify the cause of any problems arising in the program. Defining and following a well-structured testing process, with automation of regression testing, is always useful for validating the performance of an application. Good software development practices, for example, providing adequate documentation in code, and structuring the code into well-defined modules and following known patterns in code-design, are always desirable, regardless of the team size.

Following good software engineering practices will always be beneficial for code development, whether for mobile applications, for their server-based counterpart, or for general software systems.

CHAPTER 11

POWER EFFICIENCY FOR MOBILE APPLICATIONS

In this chapter, we look at the subject of making power-efficient mobile applications and the techniques an application service provider writing mobile applications can deploy to make that application more power-efficient. Battery power is a rather limited resource on the mobile phone, and writing applications that put the minimum drain on the battery is highly desirable. Given the time it takes for batteries to charge themselves, it is quite likely that users will uninstall any application that drains the battery faster than others. In that sense, writing power-efficient applications is virtually an essential requirement for any application.

Power management on mobile devices is a complex subject and many different approaches to improve power efficiency in mobile devices have been developed within the technical community [50–55]. Most of the focus on power efficiency has been on techniques that can be used within the operating system and middleware that are available on the mobile device. Application developers do not have much control on many of those techniques. However, an understanding of the approaches that are used for managing power on the devices helps in developing power-efficient applications.

For this reason, we start this chapter with an overview of power-saving techniques that can be used within a mobile device. The first item presented is a simplified model for managing resources within the mobile device

Techniques for Surviving the Mobile Data Explosion, First Edition.
Dinesh Chandra Verma and Paridhi Verma.
© 2014 The Institute of Electrical and Electronics Engineers, Inc. Published 2014 by John Wiley & Sons, Inc.

and identifying the role of the application in that model. Then, we look at some of the common techniques that can be used for improving power efficiency. Finally, we look at some of the best practices for writing power-efficient applications.

11.1 MODEL FOR POWER CONSUMPTION

The power consumption in a mobile phone comes from the use of various resources, for example, the screen, the processor, the GPS unit. As the various resources are used, they consume different amount of power and put a different amount of drain on the battery. Power efficiency requires the coordination of the state of different resources. This coordination can be modeled using a set of three abstractions.

The first abstraction is that of the physical resource itself. The physical resource is usually a hardware component in the mobile device. A resource can be turned on or off, or be put into one of several modes, each of which could use a different level of power. The resource's operations are controlled by a component in software, which is the second abstraction, that of a resource manager. The resource manager is responsible for managing the state of the resource, and specifically deciding which of the many power modes the resource ought to be put into. Normally, there will be one resource manager for each resource in the device. The third abstraction is that of the resource consumer, a software component that needs access to the resource. There could be several resource consumers in the system, and the same software component may require access to more than one resource manager (Figure 11.1).

Mobile applications that developers write would usually be resource consumers, while the resource managers would be provided by the operating system of the mobile device. The resources are usually hardware devices. Depending on the implementation of the operating system on the mobile

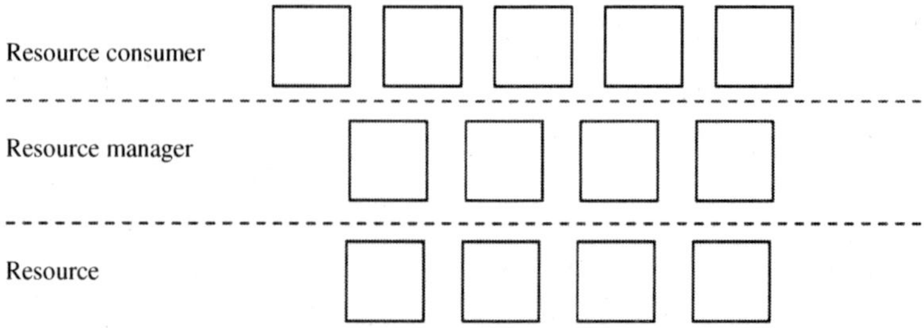

FIGURE 11.1 Simple resource model for power management.

device, the applications may have direct access to the resource manager, or that access may be hidden from the application, and the resource manager can only be indirectly invoked when the application calls some higher-level interface exposed by the operating system.

Each resource that is turned on within a mobile device consumes some amount of power. The total power drain that is put on the device overall is the combination of the power consumed by each of the resources that are currently active within the system. Therefore, an effective approach to minimize power consumption is to try to reduce the number of resources that are used by an application.

Each of the devices within the mobile device have different amount of power consumption. The exact power consumption rate would depend on the model and the usage pattern of the device. However, some resources such as the network interface, the screen, the GPS component, or the graphics card are usually big sources of power consumption [55]. At any time, the resource managers have to keep any resource that is requested by any of the applications up, resulting in battery consumption. The key to reducing battery consumption is to reduce the number of resources that are powered on.

In order to enable the optimum power savings, each resource consumer should request access to resources that they actively need, and give up the resource as soon as its function is over. This will allow the maximum flexibility to the resource managers to turn off any of the resources that are not being actively used by any of the consumers.

Once the resource consumers have made their requests to the resource managers, the resource managers can use a variety of techniques to minimize the consumption of power. Some of the common techniques used for this purpose are listed in the next few sections.

The ability of any application developer to use these techniques is impacted significantly by the nature of the device for which the application is written. In general, all of these techniques are available to the operating system developer in the system. The resource managers are a component of the operating system. However, some operating systems may not export the interfaces for requesting resources in the software development kit available to application developers. Other operating systems would offer a power management interface to application developers, but only with limited abilities to control the state of a resource. As a result, the application developers may only be able to use a subset of the different techniques that are available for power management.

In the remainder of the chapter, we have gone into more detail on techniques which the application developer can use, providing only a high-level overview of the techniques available mostly to operating system developers.

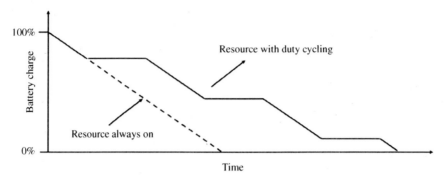

FIGURE 11.2   Duty cycling impact on battery life.

11.2 DUTY CYCLING

The concept of duty cycling is relatively straightforward. When a resource is not being used, it can be powered off. It is only turned on when there is a resource consumer actively using this resource. Duty cycling allows one to reduce the power consumed by a device when it is not actively used.

Figure 11.2 provides a simple illustration why duty cycling works. Assume that a resource consumes power at a constant rate when it is powered on. Starting from a fully charged battery, this would result in a linear discharge of the battery, and the single charge of the battery can run the resource till the battery charge runs down to zero. In real life, batteries and resources tend to have a significantly more complex relationship, but the simple model will suffice to illustrate the effectiveness of duty cycling. Now consider the charge on the battery when the resource is not used for some periods of time. This would result in the battery charge remaining the same for periods of inactivity and allow the same battery to last for a longer period of time. Figure 11.2 shows the graph when the resource is turned off during three intervals of time, allowing the battery to run the resource for a longer period of time. The battery life using duty cycling is much higher than the battery life without duty cycling.

The effectiveness of duty cycling depends on the power consumed by the resource when it is on, the power consumed by the resource when it is turned off, and the time it takes for the resource to be turned on or off. If a resource can be turned on and off relatively quickly, duty cycling is more effective. Duty cycling is less effective for resources that take a long time to turn on or off.

Suppose it takes the time t_{on} for a resource to be turned on from an off state and the time t_{off} to turn the resource off. The resource can be turned off for any time period t greater than $t_{on} + t_{off}$ where no request for the resource is anticipated. The larger the time period when the resource can be turned off, the more power savings will be attained.

The scheme to decide when to turn on a resource and when to turn it off is called duty cycling strategy, and has the most crucial role in determining the power efficiency of the device. The strategy would generally predict how long the resource is to be turned off. In the case of resources where the t_{on} period is fairly small, the strategy could be to keep the resource turned off until a request for the resource is received. When this period is not negligible, strategies to forecast when the next request for the resource will arrive are used to determine the time the resource ought to be turned off.

11.3 POWER MODE MANAGEMENT

Power mode management is a generalization of the concept of duty cycling. In duty cycling, we considered a resource to be operating in two modes, either turned on or turned off. When the resource is turned off, it is using negligible amount of power. However, some resources are more flexible and allow several different modes of operation, each mode consuming power at a different rate.

A common example of a resource using different power modes is the processor in modern mobile devices. These processors generally consist of several hardware threads, and some of these processors allow operation in modes where the clock frequency can be turned down. Depending on the number of hardware threads that are turned on, and the clock frequency selected for operation, the processor consumes different amounts of power.

In power mode management, the resource manager needs to decide which of the different modes the resource ought to be running into. The estimation of the operational mode is determined on the basis of how much requests for the resource is needed, and which mode will be the most suitable to use to satisfy those requests in any given period of time.

11.4 COMMUNICATION AND COMPUTATION CLUSTERING

Some studies of power consumption in mobile applications have indicated that most of the energy is spent in input/output to devices (e.g., storage) and such events are clustered, often called by a few frequently occurring software routines [53]. If the events that require access to a resource are clustered, then the effectiveness of techniques like power mode management or duty cycling can be improved upon.

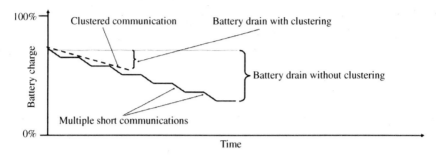

FIGURE 11.3   Clustering impact on battery life.

The concept of communication and computation clustering is to try to cluster the activities associated with a resource in a manner so that they can be satisfied in a minimum possible amount of time which is contiguous together. Suppose a mobile application wants to check for an update on a periodic basis, for example, it is an application that shows advertisements to the user, and a new advertisement needs to be shown once every second. Such advertisements, while considered an annoyance by many users, are a feature found in many freely available applications.

A simple implementation approach for supporting advertisements would be for the application to check for a new advertisement from a server once every second. Assuming this is the only communication from the device to the external network, the network communication interface resource needs to be turned on every second for this communication to happen. Suppose p_{on} is the power consumed in turning on the resource, p_{off} is the power consumed in turning off the resource after the communication is over, and p_{ad} is the power consumed in getting a single advertisement from the remote server. The device is consuming power of $p_{on}+p_{off}+p_{ad}$ every second.

On the other hand, an alternate way for the application would be to fetch a set of 10 advertisements every 10 seconds, show those 10 advertisements for the next 10 seconds in some order, and then retrieve another set of 10 more advertisements. In this case, the application is consuming the power of $p_{on}+p_{off}+p'_{ad}$ every 10 seconds where p'_{ad} is the power consumed in retrieving the set of 10 advertisements. Normally, p'_{ad} would be significantly less than 10 times the power required to get a single advertisement. If we assume that p'_{ad} was five times p_{ad}, we see that the power consumed would be $(p_{on}+p_{off})/10+p_{ad}/5$ every second. This is significantly less power consumed than trying to get an advertisement once every second.

Figure 11.3 shows how the battery power would be utilized as a result of clustering of communication analogous to the example shown earlier. The figure assumes that the battery is consumed only during the periods when communication happens. Communication that happens many times in short bursts ends up draining the battery more than if the communication happened infrequently. Even though more of battery is drained in every burst of long communication than an individual short burst of communication, clustering helps in reducing the rate of overall consumption and ends up using the power much more efficiently.

Just like communication, computational functions are also activities which invoke the processor resource on a computer. An application which can manage to do the bulk of its computations in a small burst of time and put itself to sleep for a long period of time allows more time for a processor to be turned off, thereby allowing a reduction in power consumption.

The clustering of communication and computation is a technique that application developers can use to structure their systems. The task of managing the resources on the mobile device belongs to the resource managers, which have to take the appropriate decision regarding each resource. However, each application that attempts to cluster its computation and communication provides more opportunities for the resource manager to optimize the operation of the resource.

Within the resource manager itself, clustering of activities translates to an efficient strategy for scheduling the resource usage. The resource manager can combine the requests from all of the consumers, decide on the optimal time to wake up the resource and utilize it, as well as how to batch the different requests together so that it can be performed in the most power-efficient manner by the resource itself.

11.5   EFFICIENT RESOURCE USAGE

Applications as resource consumers are the generators of the workload which drives resource usage. As applications make requests for different resources on the mobile device, they can structure themselves so that they are using the resources in an efficient and intelligent manner. The proper use of resources by the application can play a crucial role in the efficiency of the system.

Towards that goal, an application should be written so that it keeps a resource for the minimum amount of time that it is needed. If a resource is not needed immediately, it should be released so that its resource manager is free to turn down its power, if no other application is using that resource. As an example, let us consider a processor which has different memory modules, which can be turned on or off independently depending on whether they are used or not. An application which is inefficient in its memory management, and has a significant amount of unused memory that it has allocated to itself, forces those memory modules to remain on and consume power even if it serves no useful purpose. In this situation, if an application has a memory leak, in which the amount of allocated but unused memory grows over time as the application runs, the power consumption of the memory module will be at its maximum even though no useful work is getting out of the modules that are turned on. Checking for efficient memory usage and avoiding memory leaks is crucial for power-efficient mobile applications which are written in languages requiring explicit memory management (e.g., Objective C++). It will be less of an issue in programming environments where the runtime does memory management on its own and does not allow explicit memory management by applications.

In addition to only using resources that are necessary, the application developer should also take into account the utilization of the different resources that are available to it, and ask for resources so that they operate at a reasonable utilization. As an example, let us consider a multithreaded application which uses multiple threads for performing its functions. In a simple minded application, the developer may create a large number of threads, for example, up to the maximum number of concurrent threads that the underlying operating system or hardware can support. However, if many of these threads are idle, a significant amount of processor time can be spent in doing the required housekeeping to manage the large number of active threads. Instead of pre-allocating a large number of threads, the application developer may want to consider creating new threads as the workload on the application increases. This may allow the application to become more power-efficient.

The use of operating system or hardware features that provide more efficient ways to perform the same function must be considered by an application developer. If an ability to be notified of an event asynchronously is available to the application, it would be better for it to use that notification instead of waking up periodically to check for the occurrence of the event. In general, if a function is available from the operating system or underlying middleware used for application development, it will be more efficient in terms of power for an application developer to use that feature instead of implementing equivalent functions on its own.

11.6 BEST PRACTICES FOR APPLICATION POWER EFFICIENCY

Because of the importance of power management in mobile devices, many set of best practices for creating power-efficient applications can be found in the technical literature [56–58]. The best practices have been developed for a variety of reasons, some of which are orthogonal to power efficiency aspects. Nevertheless, many of the guidelines provide practical ways in which mobile applications can be written to maximize power efficiency. In this section of the chapter, we present a combined view of the various best practices, selected those among them which have a strong impact on the power efficiency of the applications.

11.6.1   Best Practice 1: Minimization of Application and Content on Devices

The resources required by an application on the device can have a significant impact on how much resource is used, which in turn drives power efficiency. Mobile applications may need a variety of content including images and data files. Since many mobile devices have limited screen sizes, large images that will not provide any perceptible increase in quality ought to be scaled down so that they are smaller. Similarly, the sizes of files that are stored and created on the local devices should be kept as small as possible. Reducing the amount of resources used on the mobile device helps in making applications power-efficient.

11.6.2   Best Practice 2: Maximum Usage of Server-Side Capabilities

The complementary aspect of the first best practice is to use as much of server-side capabilities as needed. Most mobile applications work in conjunction with a server, and moving as much function as possible to the server side works well to improve resource consumption and power efficiency on the mobile device.

An example of such exploitation of server-side capabilities is the use of assisted GPS. GPS is generally a power-hungry capability on the mobile phones. In assisted GPS [54], much of the computations needed to identify the location are moved over to a server which can perform the calculations and thereby reduce the processing and power required of the mobile phone.

Another common place where an application can use capabilities on the server is in detection of the mobile device capabilities. Given the wide range of mobile devices that are available, it is common in web-based applications to detect the type and nature of device which is running the application and then customize its operations and display appropriately. This detection can be done on the server side, for example, by using some of the fields of the web request sent to retrieve the application to identify the customizations, or it can be done when the application actually runs on the mobile device. The former will be much better from an energy efficiency perspective.

11.6.3   Best Practice 3: Batching of Network Requests

Most mobile applications need to make requests over the network to access information and resources from a server. It will save power, as well as network bandwidth in most cases, if these requests are batched together to make as few requests as possible. The batching of network requests is an example of clustering of communication discussed earlier in the chapter.

11.6.4   Best Practice 4: Adaptive Behavior of Applications

Applications take many actions that are useful and efficient during periods of high activity, but not useful in terms of getting work done when there is not enough going on within the application. In many cases, the amount of time spent in these actions can be adjusted depending on the amount of workload, and the adjustment can be made so that the system works in the most power-efficient manner.

An example of such an action is polling a server to monitor any changes that may have happened. Instead of setting a fixed period at which the application polls the server for any changes, the application can adaptively adjust the period of polling in accordance with the rate at which information changes on the server. Furthermore, the application can delegate the determination of the polling interval to the server (use of best practice 2) in order to minimize the consumption on resources on the mobile device.

Earlier in the chapter, we looked at other instances of this practice, such as an application determining the number of threads to be used on the system in a dynamic manner, instead of going with a static choice.

11.6.5   Best Practice 5: Designing for Minimum
Screen Brightness

One of the big consumers of power in modern mobile devices is the screens used in them. A significant difference in power usage can be seen in any device once the screen brightness is turned down to the lowest possible level, as compared to operating the device with the higher level of screen brightness. Applications that can function well without much impact on user quality of experience will help maximize the battery life of the system. The application designer can use a variety of techniques to achieve this objective.

If the application uses text in its user input and display that has a higher contrast with the background color, then the text will be more visible with a dim screen. Similarly, if the application has a minimum level of options displayed on any single screen during its design, each of these options can be larger, enabling better readability and visibility on a dim screen. The use of a good typeface that improves readability can also enhance the ability for users to operate the application on a dim screen.

11.6.6   Best Practice 6: Awareness of Current Device Context

Mobile applications are written to work on many different types of devices, which can be used in many different modes and manners. The context of a mobile application identifies the environment in which it is running, for example, the display screen size of the mobile device, its operating system, etc. Beyond this basic concept of context, one can envision other aspects of context that an application should be aware of. A set of definitions of context and some examples of context-aware applications can be found in Reference 58. Some of these other aspects include the awareness of the current battery level on the mobile device and the awareness of the nature of network connectivity the device has.

If an application has the ability to run in different modes, it may want to select the mode of operation depending on the battery power left in the device. As an example, an application can control the frequency at which it polls a server for any changes in information. It can make the polling interval larger if it sees the battery going down, or stop its network interactions if the power has fallen below a critical threshold. Similarly, the application can determine how many threads it should keep concurrently active taking into account the amount of battery power that is left on the device.

Another aspect of the context is the current connectivity of the device. If an application detects that it is connected via the Wi-Fi interface of a mobile device, it can go into a different mode of operation than when it detects that it is connected to a cellular network data interface only. In general, the Wi-Fi network connection provides for a higher bandwidth (and in many cases a lower cost access) than the cellular connection. In that case, the application may be able to move more of the processing functions and capabilities to a server in the network, thereby conserving power.

The exact context of operation, and how exactly it can be exploited for power efficiency, will be dependent on the nature of the application. However, designing applications which are aware of the context could help significantly in improving their power efficiency.

CHAPTER 12

BANDWIDTH EFFICIENCY FOR MOBILE APPLICATIONS

In this chapter, we look at the subject of making bandwidth-efficient mobile applications. As mentioned earlier in this book, bandwidth is a limited resource for wireless communications and is likely to remain limited for the foreseeable future for mobile applications. The application developer has the capability to control both parts of the application, the part that runs on the mobile device and the part that runs on the server backend. Therefore, the application developer is in a unique position to implement techniques that can run efficiently over bandwidth-constrained networks.

Many of the basic techniques required to write efficient applications in the presence of constrained resources are not new. These were practiced by application developers during the early stages of distributed computing systems in 1960s, an era where network bandwidth as well as processing capacity were limited. Distributed computer games were written routinely to operate well over telephone dial-up modems using extremely limited amount of bandwidth. Some of the same basic techniques need to be reused in the context of modern mobile applications to enable them to run with the least possible usage of bandwidth.

There is a financial incentive for making applications more bandwidth efficient. In many cases, mobile network operators are moving away from unlimited data plans to a usage-based data plan. In these cases, having applications that use less bandwidth will be more attractive to the end users.

Techniques for Surviving the Mobile Data Explosion, First Edition.
Dinesh Chandra Verma and Paridhi Verma.
© 2014 The Institute of Electrical and Electronics Engineers, Inc. Published 2014 by John Wiley & Sons, Inc.

Bandwidth-intensive applications would become too expensive to own due to data charges. Alternatively, their usage will be reduced since users would only operate them when connected to a free network, for example, when they are on their home Wi-Fi network. Bandwidth-intensive applications that are too expensive to own are more likely to be uninstalled by the end users.

12.1 PRELOADING

Preloading is the approach of downloading the major part of data needed for running any application at the time when the application is downloaded and installed on the device. Many networked multiuser computer games used this approach to give an exciting visual interface to users while the game operated over relatively slow dial-up links. The icons, images, and other objects in applications that could add to the bandwidth consumption were downloaded at the installation time. When the game would be invoked, only relatively small amount of data needed status change notifications, for example, position updates of different users in a simulated game world, needed to be exchanged. This reduced the amount of bandwidth needed by the applications while decreasing the delay that was involved in synchronizing the status of the games between different users.

The same technique can be used in modern applications, including but not limited to modern games. The bulk of information needed by the application can be downloaded at the installation time of the application, while only periodic small pieces of information needed by the application are exchanged with the server over the network.

In many mobile devices, such complete installation of the application may not be possible because of the limited resources available on the mobile device. In those cases, some of the other techniques described in this chapter would need to be used.

12.2 COMMUNICATION CLUSTERING

Communication clustering was discussed in the previous chapter as a technique that makes an application more power efficient. In addition to that aspect, clustering together of different network communication in one big batch has the potential advantage of making the application more bandwidth efficient.

The bandwidth efficiency arises from the fact that all network communication is associated with an overhead in terms of bytes that are exchanged. In any communication exchange, some amount of data is exchanged by the application running on the mobile device and the application component on the server. This is the useful data that needs to be exchanged. However, each exchange also requires the transfer and receipt of additional bytes that are required by the underlying protocol. These additional bytes, which are not used by the application but necessary for the communication to happen, are the overhead associated with the data exchange.

Suppose you are writing an application that needs to exchange 1 kilobytes of information every second. In order for this information to be sent, the application needs to establish a communication session between the mobile device and the server backend. This establishment requires the exchange of a few overhead bytes. Once the session is established, the 1 kilobyte of useful data needs to be wrapped into the protocol data units which have their own additional headers which are also part of the overhead bytes that are transferred.

Assume now that you can cluster the information exchange so that the application exchanges 10 kilobytes of information every 10 seconds. The overhead bytes would remain the same, except that they apply to 10 times the amount of useful application level data. This reduces the net amount of bytes exchanged over the network, while the same amount of useful data bytes are exchanged among the applications.

The extent to which communications can be clustered depends on the nature of the application and its communication needs. If the data being exchanged by the application is not time-sensitive, it can be clustered easily without impacting operations. On the other hand, if a piece of data needs to be retrieved within a specific time deadline, it would constrain how much communication clustering can be done.

Another technique that can be used in conjunction with communication clustering is throttling. Communication clustering may require a lot of communication to occur in a burst at the same time. Throttling puts a limit on how much network communication rate will be supported by the application. This limit can be used to prioritize which network communication requests ought to be sent at any time.

12.3   CONTEXT-AWARE COMMUNICATION

Until now in this chapter, we have considered all network communication to be equivalent. However, not all types of bandwidth exchanged by the application are equivalent in terms of the cost to the user or in terms of the speed at which the information can be exchanged. It frequently depends on the context of the communication, where context can be defined as the environment in which an application is present, including its location, current environment, and use of the application [59,60]. As an example, for a user with personal

Wi-Fi access at home, any information exchanged while connected to the home network would come free of cost, while a usage-based fee may be associated with the exchange of bytes on the cellular data interface. The nature of connectivity in this case defines the context of this information exchange. Being aware of the context, that is, the current connectivity can help an application make intelligent choices about optimizing bandwidth usage. The concept of context is more general than just the type of network connectivity. Context can include any attribute of the user, the device or the current state of the device, for example, the location of a user, the motion trajectory of the user, and the time when a piece of information is required to be delivered to the user.

Applications that are aware of the context of communication would be able to optimize the bandwidth exchange to fit within a desired monetary or power budget. Context-aware communication allows the application to get information delivered asynchronously, that is, at a time when the network is able to deliver it instead of when it is requested by the user. Several approaches for context-aware communication based on asynchronous delivery have been proposed in academic literature [60–62] although they have not yet found wide adoption in the commercial arena.

12.4 DISCONNECTED OPERATION

Some mobile applications can be written so that they can allow a user to operate even when the mobile device is not connected to the server in the backend. As an example, some email applications allow users to compose and write email to other users even when they are disconnected. The outgoing emails are held at the mobile device until the network connectivity is established and then transferred over to the server side for subsequent delivery.

Applications that can provide a disconnected mode of operation can take advantage of holding all of the network communication tasks until the network connectivity is established and then performing all of them in one big cluster. In essence, the transfer of the application mode from disconnected to connected mode is the trigger point for batching of requests for communication.

Disconnected operations can be conducted by the user manually deciding which of the modes the application should run in, connected to the network or in a disconnected mode. However, this determination of the mode to run an application can also be done automatically by the system, for example, by determining the context of the mobile device and deciding on a set of predefined rules whether the connected mode or disconnected mode ought to be used for the application at any given time. When the mode of operation switches from disconnected to connected, the clustered communication requests are transmitted on the network. As long as the application supports both the disconnected and connected mode of operation, it can optimize the network communications for itself at selected times during the transition between different modes.

12.5 CACHING

Caching content at the mobile device is an effective approach to reduce the amount of bandwidth needed over the network for some classes of applications. While caching, the application on the mobile device stores some of the data that may be frequently accessed in a local cache in the application itself.

Caching on the mobile device is similar in concept to caching at intermediate proxies but has a different set of characteristics driving its effectiveness. The cache in an intermediary proxy is accessed by many different mobile device users, and the larger the number of users sharing a cache, the higher the probability that the same resource will be accessed again, thereby increasing the cache hit probabilities. On the other hand, the cache for an individual mobile user will only be used by a single person, so the gains in caching performance resulting from large number of users are not realized.

Application writers however have more knowledge about the types of access pattern that may be expected for their users, and this knowledge allows them to make a better determination of the content that is cacheable. While intermediary caching proxies are likely to be used for application protocols that are common, application level caching can be used with protocols that are less common. Similarly, application flow that happens on secure protocols, for example, Secure Socket Layer or Transport Layer Security, cannot be cached by an intermediary in the network but can be cached by the application at the mobile device which decrypts the content.

The effectiveness of caching on the mobile device is highly dependent on the nature and design of the application. However, it is an useful tool that can be very effective for some classes of applications.

12.6 COMPRESSION

Compression of the data that needs to be exchanged across the network is a technique that is very effective for use within the network, as well as between the application and the server components of a mobile application. The application can use insights about its behavior to determine the right nature and type of compression that can be used.

Application-specific compression needs to be done on the content before it is transmitted on the network. Since the bulk of transmission on the network happens to be from the server to the mobile device, the onus of compression falls largely on the server side of the mobile application. This works well for mobile devices which are often limited in power and resources. The nature of most compression algorithms is such that the process for content compression consumes significantly more processor and power than the process for content decompression.

Some applications can make application-specific optimizations that can provide for a better compression that is not possible otherwise. As an example, the application can decide whether it can use lossy compression for part of the data being exchanged. Compression techniques can be divided into lossy compression and lossless compression techniques. In lossless compression techniques, the uncompressed content can be recreated with perfect fidelity from the compressed content. In lossy content, the result of the decompression does not result in a perfect copy of the original but can have some losses. Video compression, for example, uses lossy compression to reduce the size, since some of the finer patterns in video contents can be lost without any changes in perceptible quality. Lossy compression generally results in much better compression ratios than lossless compression. The application has enough awareness about the content to determine which compression technique can be used in which specific instance.

The application can also do special types of compression which can tradeoff some of the network communication bandwidth with additional processing on the device itself. One example from networked multiplayer computer games is the use of dead reckoning. Such games typically need to keep track of the position of each of the different players in the game. However, instead of getting the exact location of all the other players in the same region as its own user, the application can extrapolate the anticipated position of other players on its own without checking with the backend server, which can now be contacted much more infrequently. Only the application developer is in a situation to determine whether or not special techniques like dead reckoning are appropriate for its environment.

12.7   CONTROL TRAFFIC IMPLICATIONS

Because mobile applications send their data on cellular networks, they can inadvertently create a traffic congestion issue within the cellular network even as they are implementing techniques to save on the overall network data bandwidth. Within cellular networks, connections between a mobile device and Internet are established using a set of protocol exchanges which are classified as control traffic. When a mobile device powers on or powers off its network interface, it causes an exchange of control traffic to occur which sets up or tears down the connection. Normally, mobile devices would power down the network interface if they find that there are no active connections transferring data across the network.

Depending on how the application behaves, it can cause a flurry of control traffic activity on the cellular network which can put a significant strain on the network radio resources and the power consumption of the radio network interface on the mobile device. Studies conducted on some popular applications [63] show several instances of application behavior that can cause unnecessary network control traffic. These include the periodic retrieval of information with very small amount of data, establishing multiple network connections to get different set of files instead of getting multiple files on the same connection and creating a lot of small bursty requests. The same study also found that maintaining a constant streaming rate of data at a relatively low bit rate is inefficient from power and control traffic perspective. Network connections that are maintained open but only retrieve small bit rate of data have a much higher ratio of control traffic to useful data content and thus waste resources.

Adapting application behavior to reduce the amount of control traffic generated per useful byte of information exchanged at the application level can reduce the power consumption of an application. In the next section, we look at the set of best practices that an application can follow to become more bandwidth efficient, both from the perspective of data traffic as well as control traffic.

12.8   BEST PRACTICES FOR BANDWIDTH EFFICIENCY

As mentioned in the previous chapter, there have been several best practices proposed for developing mobile applications [56–58]. In this section, we present a combined view of the various best practices that deal with improving the bandwidth efficiency of the different applications, combining these best practices from the insights gained in application profiling studies such as "Profiling resource usage for mobile applications: a cross-layer approach" [63].

12.8.1   Best Practice 1: Provide User Control of Network Bandwidth Usage

Network bandwidth communication is not free in many cases, and some of the application's network bandwidth usage can prove expensive for users in some situations. As an example, international data roaming rates are fairly expensive compared to data rates in one's domestic calling area. If an application performs an operation like retrieving version updates when a user is travelling beyond their domestic plan areas, it can cause a significant expense to the user. Intuitive policies that allow the user to control when network communication happens from the application would allow the users to not be sprung with surprises due to unexpected network bandwidth consumption.

12.8.2   Best Practice 2: Minimize External Resources

Network communication could be avoided if there are no external resources that the application needs to access. Therefore, minimizing the amount of resources that need to be accessed beyond the mobile device will help reduce network bandwidth communication. When designing the application, one can structure it so that the bulk of the resources are stored on the mobile device during the application installation process. In that case, only relatively small amount of network exchanges need to happen during the runtime of the application.

The practice of minimizing external resources does have an undesirable side effect in that it can increase the amount of resources consumed at the mobile device. During the design of the application, the tradeoff in network bandwidth versus device resource consumption need to be taken into account to determine the right mix between device resources and networked resources.

12.8.3   Best Practice 3: Make Fewer Large Requests

Network communication is more efficient, both in terms of reducing the number of overhead bytes and the amount of control traffic, if the application makes large amounts of data transfer at the full peak network capacity. Making a lot of short network requests, transferring a small amount of data, or having a long running request which only uses a small fraction of bandwidth capacity creates a much larger number of overhead bytes than is necessary.

Communication clustering to make several requests together, batching requests that are not delay-sensitive, and retrieving large amounts of data in one request make network communication more efficient. Streaming applications that require a continuous feed are better off retrieving the information in periodic bursts at high speeds that are spaced out infrequently.

12.8.4   Best Practice 4: Combine Multiple Data Objects into a Single Large One

When several network resources need to be retrieved from the server side, it is more efficient to combine them into a single big object for the purpose of retrieval. Instead of retrieving each of the objects individually, the application should just retrieve the larger object in one go. The combined object may be kept as a single object at the server side, and the client can create individual resources from it after retrieving it. If there are multiple images needed for an application, it will be better to combine them together into a single file from which the client extracts them. Some news sites display a scrolling thumbnail of images. It will be more efficient for them to retrieve the data for all the images in one go rather than trying to obtain them one image at a time.

Another advantage of combining objects into a single one comes from the nature of downloading the objects to the mobile device. It will be inefficient for the mobile device to open a separate TCP connection for each of the objects and then retrieve the objects one by one on a new TCP connection. It will be more efficient if the mobile device retrieves all of them on the same TCP connection.

12.8.5   Best Practice 5: Avoid Inefficient Redirections

Some types of network interactions between the mobile device and the server are relatively inefficient in the sense of causing short low-data exchanges in a bursty manner. An example of this is HTTP redirection. Many mobile applications written in the manner of web applications would tend to use this feature of HTTP protocol to decide the right place for a client to connect. However, such redirection causes the clients to establish new connections, which causes additional overhead data traffic, as well as additional control traffic on the cellular network. Instead of using HTTP redirection to request a mobile device to connect to another website, the application developer can implement a proxy at the server side which retrieves the information from the desired server and then sends the results to the client. This will save both power and network bandwidth at the client.

12.8.6   Best Practice 6: Compress and Minimize the Data Being Exchanged

Whenever there is a need to exchange data over the network, it is a good idea to compress and minimize the amount of useful data to the extent possible. In many cases of web-based applications, text in the content that is put in for usability and readability (e.g., comments or white spaces) can be removed if they are going to be processed automatically on the mobile device. Similarly, careful design of application structure can reduce the amount of information that needs to be exchanged. Data objects can be compressed and video content down-sampled according to the characteristics of the mobile device accessing the device.

12.8.7   Best Practice 7: Diligent Use of Cookies

Cookies are frequently used in web-based applications to store information on the mobile device that is personalized for the user. The information stored in the cookies can become very large if cookies are not managed properly. While it is not an issue on laptops or personal computers that have good connectivity and ample resources, it can cause quite a drain on the network capacity when bandwidth is at a premium. Instead of storing large cookies at the client, the better option will be to have a cookie store on the server side. The cookie on the client can be a small index that looks up the appropriate entry on the server side to obtain more detailed information.

Cookies typically allow specification of which domains and sites they can be exchanged with. When dealing with mobile devices with constrained networks, it is best to make the sharing be as restrictive as possible. In that way, minimum number of cookies is sent to any site that is being accessed and unnecessary exchange of cookies is avoided.

12.8.8   Best Practice 8: Use Intelligent Caching

Caching for content that can be used frequently is a good way to bypass the need to access the content over the network. Resources that are going to be used frequently by an application can be cached at the mobile device to avoid the need to retrieve them repeatedly from the network. The application can intelligently decide which elements to cache locally to maximize user experience and minimize network communication.

Sometimes the same object can be accessed with different names. If the application is aware of such type of aliasing in the names of objects being accessed, it should use fingerprinting to decide if an object is in the cache, rather than using just the name of the object. A fingerprint could be computing a hash of the contents of the object and comparing it against the hash of objects stored in the cache to determine if the object is in the cache. Trying to maintain a fingerprint of the content can improve the effectiveness of caching.

12.8.9   Best Practice 9: Context-Aware Communication

A context is the environment under which the mobile device is operating, including aspects like its location, its current network connectivity, its current power level, and its current usage pattern. Understanding the context of a mobile device and using that to control network communication can result in significant savings of bandwidth.

If the application sees that it is connected to a network that has a limited bandwidth capacity, it can switch to a lower resolution model for communication or switch to a disconnected mode of operation. Similarly, the application can make different choices for how to behave when it has connectivity via a Wi-Fi medium, connectivity via the cellular network interface in the domestic calling area, or connectivity via the cellular network interface when roaming outside the domestic area.

Similarly, the application can determine whether it is being used actively by the user or not. If the user is actively using the application, the application may contact the backend server for any required information more frequently than when it is not being used actively. If the user has just left the application on without invoking it, the application may want to prolong the polling interval to save on network bandwidth. Adjusting the behavior according to the activity or inactivity of the user (which is a part of the extended concept of context) can make an application more bandwidth efficient.

CHAPTER 13

MOBILE DATA ISSUES FOR THE ENTERPRISE

The enterprise makes up an important segment of the mobile data ecosystem. Enterprise users are consumers of mobile devices, and enterprise IT departments are responsible for development and deployment of many applications on the mobile devices geared towards their employees and customers. The growth of mobile devices has had a transformative effect on many enterprises, allowing them to do many functions in a new cost-efficient manner. Any discussion of mobile would be incomplete without a discussion about the impact of mobile data on the enterprise.

Most of the chapters in this book have dealt with the issue of growth of mobile data, dealing with the communication challenges associated with mobile network traffic and the opportunities created by the growth of mobile devices. The enterprise users form a special case where the growth in data is not a significant issue for them. Other issues, such as security and managing the diversity of devices, are much more of a concern for the enterprise.

Mobile devices operate using a wireless interface, which could be either using a cellular (UMTS/LTE/CDMA, etc.) or Wi-Fi (802.11) technology. In most modern devices, both types of communication interfaces are supported. Wi-Fi can support a higher network bandwidth than the cellular data networks, and it is fairly common for users to switch to Wi-Fi networks whenever it is available. A nontrivial incentive of this tendency to switch may be that data

Techniques for Surviving the Mobile Data Explosion, First Edition.
Dinesh Chandra Verma and Paridhi Verma.
© 2014 The Institute of Electrical and Electronics Engineers, Inc. Published 2014 by John Wiley & Sons, Inc.

transmitted on Wi-Fi networks are usually not counted as part of the data plan by most carriers. Within most enterprises, Wi-Fi networks are ubiquitous, and employees of the enterprise switch over to Wi-Fi networks when they are on premises.

Because of the use of Wi-Fi networks, mobile data growth has not been a substantial concern for enterprises. While the growth of mobile devices does require that the Wi-Fi networks in the enterprise have sufficient capacity, the upgrade of the network infrastructure by introducing additional access points is not a significant burden on most enterprise IT budgets. However, there are several other issues that are much more pressing for the enterprise users.

We begin this chapter by enumerating some of the concerns that the introduction of mobile devices causes for the enterprise. This is followed by a set of techniques that can address each of those concerns.

13.1   MOBILE-RELATED ISSUES FOR THE ENTERPRISE

Perhaps the most significant concern caused by the use of mobile devices in an enterprise is that of data security. Enterprises are formed normally with a commercial purpose, for example, manufacturing of computer or consumer products, production of goods and services, or other types of economic activities. In almost every enterprise, there is information that would be considered sensitive, namely information that needs to be kept securely and not fall into unauthorized hands. As mobile devices are more susceptible to being lost, access to the device may fall in wrong hands. Enterprises need to have sufficient measures in place with mobile devices to manage security of sensitive information despite the mobile device getting into wrong hands.

Many enterprises are establishments that have been around for a long time. As a result, they have a substantial IT infrastructure present already in their environment to conduct their business. As mobile devices become ubiquitous, the enterprises need to plan out how they would make their existing software applications and services available to the mobile device users. The enablement of mobile devices to access existing applications is another big issue that enterprises face with mobile data.

The introduction of a large number of mobile devices can put a strain on the networking infrastructure of an enterprise. Techniques to deal with these strains are also of relevance to enterprise customers. The infrastructure challenges include ensuring adequate network connectivity for mobile devices and proper management of the large number of applications and types of mobile devices that are proliferating in the mobile space. Enterprises often have their IT department manage and support devices, including mobile devices, for their employees. The management of these infrastructure challenges is an important aspect of mobile devices.

The growth and ubiquity of the mobile device has the potential to disrupt the existing business processes used by many enterprises. By addressing the issues associated with mobile adoption in the enterprise, the business benefits of this disruptive transformation can be attained.

13.2 SECURITY ISSUES

The security issues associated with use of mobile devices in an enterprise environment can be divided into two broad categories: infrastructure security and data security. Infrastructure security relates to issues about securing a mobile device or its access to the corporate resources while data security relates to issues about securing the enterprise data that is contained in the mobile device.

Security measures of any kind are never foolproof. They just make it harder for an attacker or bad person to do things that the enterprise may not want them to do. Almost any security mechanism can be broken by an attacker with ample resources, where resources include computation capability as well as smart people working for the attacker. All security measures are expensive in terms of computation, battery usage on the mobile device, and network bandwidth consumed. For any enterprise, the right security measure is one which reduces the risk due to an exposure to a level which is acceptable for the business value delivered by mobile device usage. The security measures described in this section should be considered as some of the ways to reduce the exposure but should not be considered a recipe for absolute security.

13.2.1 Infrastructure Security

Infrastructure security includes measures to secure the device as well as access of the device to the servers and applications within the enterprise network. The security of devices includes checking that the applications that are installed within the mobile device do not include malware as well as approaches to securely access the intranet of the enterprise from the mobile device. Malware is any software that is not desired on the device and may perform a variety of harmful functions like consuming unnecessary resources, capturing passwords, credit card details or other sensitive information and sending them to an unauthorized person, or launch attacks on other computers on the network. The issues in infrastructure security associated with the introduction of mobile devices are not new. The same issues were associated with the use of laptop computers or enabling remote access to the enterprise from a home computer. The only difference is the expansion of the scope to new types of devices with new operating systems.

The first challenge in infrastructure security is making sure that the mobile devices are secure, in the sense that the applications that are installed on the mobile devices do not contain viruses or other types of malware. Viruses are commonly associated with personal computers and laptops. They are less of a threat on mobile phones and tablets. The diversity in the types of mobile devices helps in reducing the threat posed by a single virus. However, mobile devices are not immune to the threat of malware. There have been more than 40 known instances of mobile phone viruses in popular mobile device platforms [64]. As mobile devices become the dominant means of user access, one would expect more viruses to emerge targeting them.

Operating systems used in mobile devices have several built in mechanisms to improve the security of mobile applications, for example, see Reference 65 for a survey of security mechanisms used in Android™. Nevertheless, some opportunities for attack remain in any complex piece of software. Malware tries to exploit such opportunities, for example, any bugs that may remain in the implementation of the system. In addition to the normal operating systems security mechanisms, mobile devices commonly use the concept of user permission to prevent the spread of malware. During the installation process, the application asks the user for permissions to access various resources. However, since application developers are not always diligent in setting the required permissions, opting to ask for more permissions than the minimum necessary for their operation, and users tend to give permissions in a relatively indiscriminate manner, the permission mechanism only has limited success in preventing the spread of malware.

The other mechanism used by operating system manufacturers is to use an extensive review process to vet out any suspicious applications before they are allowed to be downloaded from the application store. The review process is effective in eliminating many malicious programs. However, some malicious programs could still sneak in the application stores and pass the various checks performed during the review process.

Like antivirus software for personal computers, there are various types of antivirus software available for mobile devices. They look for the existence of malware, and some of them check for security vulnerabilities that may expose an application to potential malware. Not all of the after-market antivirus solutions work effectively, since operating systems restrictions on operating systems like Android make it difficult for after-market scanners to check for problems in other applications [66]. However, applications that check for potential patterns of vulnerabilities or validate if the application permissions are granting too much privileges can provide useful guidance to potential vulnerabilities.

Enterprises can reduce their exposure to unsafe applications by defining policies and guidelines on a set of allowable applications that can be installed safely on the mobile devices that are available to their users. They can also use application scanning systems to check for potential vulnerabilities.

The other aspect of infrastructure security is securing access from the mobile device to the enterprise infrastructure. In this context, the overall situation is similar to that of remote access from home computers. The secure access from mobile devices is usually provided by means of a secure Virtual Private Network (VPN) that is established from the mobile phone to the intranet of the enterprise. The VPN provides a secure way to access the enterprise infrastructure. In that respect, VPN access for mobile devices is indistinguishable from the VPN access from laptops.

13.2.2 Data Security

There are many benefits of using mobile devices in conducting the day-to-day business of an enterprise. Business can be conducted on the move, even when one is travelling, and from any location where network connectivity is available. However, the user of mobile devices also increases the risk of potentially putting sensitive data in the wrong hands.

Every enterprise contains some sensitive data that should only be available to its employees with proper authorization. The exact nature of the sensitive information would depend on the enterprise. Some examples of sensitive data include credit card number of customers, social security number of employees, banking information for employees from payroll, the design specifications of an upcoming product, and financial reports that have not yet been released publicly. In any enterprise, this information is kept under relatively tight security. The information is usually accessible only if a device is physically on the enterprise network or if the device has a VPN access to the enterprise network.

If the device is a mobile and small in size, it is more likely to be lost, misplaced, or stolen. Once the device falls into wrong hands, any information stored on the device may become compromised. The resulting loss of information can be crippling for the enterprise, depending on the nature of the compromised information. The loss of sensitive data can expose an enterprise to severe financial and compliance risks. A mobile device left with plans for future products or announcements can reveal business secrets to competitors. A mobile device lost with information about patients can cause significant liability and negative publicity for a hospital or a medical establishment. A lost mobile device with sensitive credit information of customers can cause severe embarrassment with major business impact on a financial company. As a general rule of thumb, an enterprise cannot afford to lose a device with business-sensitive data on it. One can even make the case that the sole justification for infrastructure security is to prevent the potential for a data security breach.

There are basically two types of threat situations related to sensitive data when it is accessed via a mobile device. The first threat situation is due to the existence of malware on the mobile device. If the mobile device is connected to the enterprise intranet, then the malware can possibly retrieve some of the sensitive information or be able to intercept the sensitive information that is downloaded on the mobile device. The other situation is the loss of the device. The device may fall in wrong hands because it is misplaced and someone finds it, or it may be purposefully stolen by a thief. In either case, the person with physical possession of the device has the opportunity to get access to the sensitive data that is stored on the device.

Let us look at some of the ways to protect sensitive enterprise data when the device has some potential malware on it. Malware is more likely to be downloaded on a device if it uses applications that have not been reviewed and analyzed by the enterprise IT team. It frequently happens if a device is used both for personal use and business use. During personal use, people may download games, video, or other type of applications where the possibility of a malware sneaking in undetected is much higher. Business applications that are reviewed and pass the security checks of the enterprise IT department are less likely to have such malware. One way to reduce the risk is to have two devices, one used exclusively for business and the other used exclusively for personal use. If an individual has two separate devices, and the one dedicated for business is only containing applications vetted and approved by the enterprise IT department, it is not likely to get any malware. Accessing sensitive data on such a device should be relatively risk-free.

The issue involved in this approach is the inconvenience of having two devices and the associated cost of the same. In practice, many people will prefer to use the same device both for business and personal use. There are a few technical approaches available to address that problem.

The first approach is shown in Figure 13.1, which is to attach an auxiliary device to the mobile device. This auxiliary device or attachment can be accessed by an application on the phone itself. However, the hardware device itself is a miniaturized computer in its own right and capable of running applications. It will simply use the phone interface as a conduit to send data over to the enterprise servers in an encrypted manner. Sensitive information is always in the auxiliary device. The auxiliary device can be much smaller than the actual mobile device and thus be more convenient to carry around. However, it is still a second physical device.

The second approach is to have a mobile device hypervisor. A hypervisor allows the running of two or more completely separate operating systems on the same hardware. The hypervisor will maintain complete isolation between a virtual business phone and a virtual personal phone on the same physical hardware phone. This approach is shown in Figure 13.2. Apart from

Auxiliary device

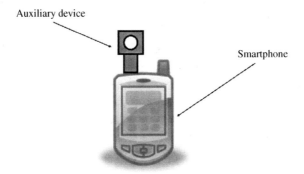

Smartphone

FIGURE 13.1   Auxiliary device approach for security.

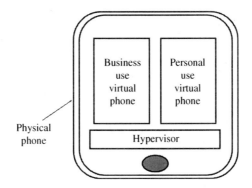

Physical
phone

Business
use
virtual
phone

Personal
use
virtual
phone

Hypervisor

FIGURE 13.2   Hypervisor approach for security.

a compromised hypervisor, there is no way for a malware on the personal virtual phone to get sensitive data that is only available in the unencrypted format on the business virtual phone. A hypervisor-based solution will work well in most commercial contexts. The only challenge is that some of the mobile phone makers may not support a hypervisor on their devices.

A variation on the theme of the hypervisor will be for the operating system to offer two different containers for applications to run in, one container for enterprise applications and one container for personal applications. An appropriately enhanced operating system can support separation between two categories of applications and allow two virtual devices on a single physical hardware. However, at the time of writing this book, none of the leading operating systems for mobile phone were offering such a system.

The third approach will be to have special enterprise applications that keep any sensitive information stored in the device in an encrypted format. The information is accessed by a password which is only known to the authorized user. A malware will need to intercept the visual display or the keyboard in order to intercept the password or the unencrypted sensitive information.

The storage of sensitive information increases the ways in which the information can be accessed but provides more security.

The fourth security approach is for the enterprise to provide an application which is just a remote display to an application running on the enterprise servers. The remote display will provide a way for the user to interact with the application on the enterprise servers. However, the information never leaves the enterprise intranet, except as transient video or graphics on the screen of the mobile device. This keeps the information secure, barring the presence of a malware that can do screen-scraping. However, the mobile application can only be used for enterprise applications when one is physically connected to the enterprise intranet. This also puts a significant load on the network because it exchanges a lot more bandwidth with the servers in the enterprise intranet. This approach is shown in Figure 13.3.

The right security approach to use will need to be determined by the enterprise as a trade-off exercise between the damages caused due to loss of sensitive information and the cost of implementing the security solution.

Let us now examine the second threat situation, that of a lost phone. Although the lost phone will likely have a password locking it, it is likely that a diligent attacker would be able to guess the password and access local data on the phone. There are various security mechanisms that an enterprise can use to counter this threat. Some of the mechanisms used for the previous threat can also be used in this situation.

Let us examine the efficacy of the previous mechanisms for this type of threat. Having a separate phone or auxiliary hardware device does not mitigate the risk since these are as likely to be lost as the primary personal phone itself. Similarly, the hypervisor solution alone is not effective in this case, since the attacker has access to both virtualized phones.

Keeping the sensitive information on the enterprise premises accessible only via a remote access protocol is effective since access can be cut off from the enterprise once the loss of the phone has been reported to the enterprise. There is a small window between the loss of the phone and the enterprise cutting off remote access to the device during which sensitive information is

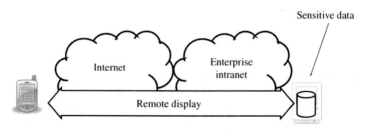

FIGURE 13.3 Remote display approach for security.

at risk. If the credentials used to remotely access the enterprise are not stored on the device itself, requiring another set of attempts by the attacker to determine those credentials, the threat in this window of opportunity can be minimized enough to be acceptable in most commercial scenarios.

Keeping sensitive information on the phone in an encrypted format helps in this case as long as the attacker is not able to break into the encrypted format. Since the attacker has a larger time in which to attempt this decryption, techniques are needed to prevent this from happening.

The first technique is that of wiping the device after a number of wrong password attempts. If the device is locked by a password, this approach allows a small number of attempts for the user to enter wrong passwords. If several wrong passwords are entered, for example, more than 10 unsuccessful attempts are made to login, the device deletes all the information contained on it, resetting it to the configuration of a new device. This allows a legitimate user to reuse the device in case of a lost or forgotten password.

One drawback in this approach is that it allows someone who has stolen the device to get access to a brand new phone. In order to reduce that possibility, the default configuration of the new device can include a location-tracking capability for the phone. This allows each phone to report its status to an asset management system of the enterprise. The enterprise can then detect that a stolen phone is being reused in an unauthorized manner and alert appropriate law-enforcement agencies to take measures against the thief.

The second technique to provide further protection to sensitive data stored in the device is to split the sensitive data so that it can only be unencrypted if information is collected from three different places: (i) some input from the authorized user not stored in the device, (ii) some data stored on the device itself, and (iii) some data that is obtainable from an enterprise server which is attainable only when the device is connected to the enterprise intranet. A typical way to do it would be to encrypt the sensitive data and then split it into two portions, one accessible from the phone and the other accessible only from the enterprise intranet, with the input from user taking the form of a short pass-code. This technique of secret sharing [67] can be applied to the encrypted sensitive data as well as the encryption keys to retrieve the sensitive data which themselves can be split in this manner to provide additional security. The secret sharing reduces the time window within which the attacker can try to get to the sensitive data to the interval between the loss of the phone and the cutting off of the lost device from intranet access. Secrete sharing can also be used to reduce the dependency on network connectivity by enabling access to encrypted data for some period of time in disconnected mode after the part available from the enterprise intranet has been retrieved.

By using a combination of these security mechanisms, the enterprise can reduce the risks in using mobile devices to an acceptable level to reap the benefits from increased use of mobile devices in business.

13.3 BACKWARD COMPATIBILITY

Apart from the enterprises that are being formed right now, most enterprises have been around since the period where most of these services were provided to remote users over the web. However, as mobile devices have become more popular and offer more capabilities, many of these services and applications are now being accessed via mobile devices. While new applications can be developed completely with a primary focus on access from mobile devices, the existing web-based applications may not be suitable for accessing from a mobile device.

There are many reasons why an existing web-based interface would not be suitable for a user accessing the same service from a mobile device [68]. Even though modern mobile devices are equipped with browsers, the browsers on the mobile device usually have two challenges: (i) a much smaller screen size than regular laptops and personal computers and (ii) a much limited network bandwidth when connected to the cellular network. The small screen size is especially tricky issue since the ability to move around with a mobile device increases as its size is reduced, yet the usability of the device increases as the size of the screen is increased. Mobile devices also have differences in interacting with users than normal computers, which can cause more challenges in user interactions with the websites [69].

The small screen size makes some web pages difficult to read on a mobile device. A display banner on top or a navigation menu on the side of a web page, a staple feature of many websites, can take up too much of real estate on the screen of a mobile phone leaving too little space for useful content. If a web page is not rendered properly, the user may be forced to scroll down several screens before getting the main content of the page. A web-based input form that relies heavily on text input may be inconvenient for mobile phone owners with smaller screens where typing in text is more tedious and difficult. Even simple matters, such as requiring a password with a mix of numbers, letters, and nonalphanumeric, characters can become tedious in a mobile device without a full keyboard as users need to switch across different split virtual keyboards to enter the required input.

One can upgrade existing websites and web-based systems so that they become more suitable for accessing via a mobile device. The only issue associated with such an upgrade is the associated costs. Any medium-sized enterprise is likely to have scores of websites of various natures and upgrading them all to become more mobile accessible has a substantial investment cost.

If only some of the websites are being upgraded for mobile device access due to cost reason, one can put a web proxy in front of the regular websites. This proxy would determine the type of device that is trying to access the website and then redirect it to either the original website (if the device is a laptop or personal computer) or to the mobile version of the website (if the device is a mobile one). The determination of the device can usually be made by the description of the browser and user-agent that is included as standard headers in a web page request. This type of redirection mechanism has the benefit that more of the existing websites can be gradually made mobile-enabled over time in a seamless manner to the existing user set. The physical setup for this approach for backward compatibility is shown in Figure 13.4.

Another option for reducing the expense of website enablement for mobile applications is to introduce a proxy that transforms the content to be more suitable for mobile applications. This setup is shown in Figure 13.5. The transformation can be done by means of set of rules that are applied generically across several web pages or by writing specific code modules that are used for transforming content belonging to specific websites. When special code modules are written to do the transformation, these modules are sometimes referred to as microapplications.

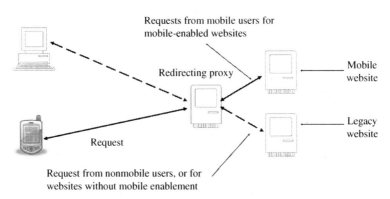

FIGURE 13.4   Redirecting proxy for backward compatibility.

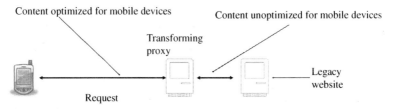

FIGURE 13.5   Transforming proxy for backward compatibility.

Both of these techniques can be used as transition strategies for making existing applications ready for mobile devices within an enterprise.

13.4   INFRASTRUCTURE ISSUES

As mobile devices grow in the enterprises, there are some infrastructure issues that need to be addressed. These infrastructure issues include the need to have sufficient network connectivity, as well as getting the challenges associated with application and device management for mobile devices.

On enterprise premises, the growth of large number of mobile devices may put some stress on the wireless networks, which may be more pronounced for sites that were not designed to enable mobile access in general. In these cases, a redesign of the network infrastructure may need to be done in order to ensure sufficient coverage and bandwidth for mobile devices to communicate with enterprise services and servers. In some cases, it may require upgrading the bandwidth of the link connecting the campuses of the enterprise. For example, if a branch of the enterprise does not have sufficient access link bandwidth to support the set of new devices that customers are bringing to the work, they may need to upgrade that link to higher capacity.

The mobile devices used in the enterprise could be owned by the enterprise or they could be personal devices belonging to the users. Some enterprises are allowing their employees to bring their own personal devices to do their work, referred sometimes as BYOD or bring your own device. When the mobile devices are owned by the enterprise, the enterprise needs to put into place a system that can track those devices, determine the right set of applications for them, and allow employees to get access to that set of applications. The enterprise would also need to have sufficient IT staff that can install those applications on the mobile devices, help employees who are having problems using the applications or the devices, help customers upgrade the applications when needed, and determine policies governing the use of mobile devices and applications.

Similar issues need to be decided in case of BYOD mobile applications. The enterprise IT department may need to define policies on when the use of personal devices is allowed, the right mobile security approach to use on those devices, and to help users install applications needed for their business functions on their personal mobile devices.

In some enterprises, customized mobile application development may need to be done. These would require a development platform, selecting a set of standardized libraries to support mobile applications, and appropriate tools for debugging, testing, and deploying mobile applications.

The collection of various mobile application development platforms, mobile application management systems, and mobile device management systems are referred to as Mobile Enterprise Application Platform (MEAP) [70]. MEAP is a loosely defined term that encompasses an assortment of tools and technologies pertaining to the infrastructure needed to support mobile applications. Given the wide set of choices that are available for mobile devices, mobile applications, development platforms, and development technologies, each enterprise needs to determine the right MEAP that is most appropriate for its own business environment.

CHAPTER 14

RELATED TOPICS

In this final chapter of the book, we look at some of the issues that are peripherally, but not directly, relevant to the topic of mobile data growth. These subjects are reviewed here briefly with references to other texts that can provide more insights into the challenges and issues associated with them.

Topics related to the subject are Machine-to-Machine (M2M) communications, the Internet of Things, participatory sensing, the impact of mobile devices on business, Software-Defined Networks (SDN), the mobile first philosophy, and network analytics.

14.1 MACHINE-TO-MACHINE COMMUNICATIONS

Mobile data growth has been driven largely by the use of mobile phones and tablets. These mobile devices are used by humans to access services over the Internet via wireless networks. The communication is thus between an individual and a computer.

However, instead of having a human at one point of the communication, one can envision that both points of the communication are computers. In this case, the communication is between two machines. The set of applications

Techniques for Surviving the Mobile Data Explosion, First Edition.
Dinesh Chandra Verma and Paridhi Verma.
© 2014 The Institute of Electrical and Electronics Engineers, Inc. Published 2014 by John Wiley & Sons, Inc.

which require the communication between two machines comes under the umbrella of M2M communications.

There are several examples of M2M communications. These include the monitoring and control of electric consumption in households creating a smart grid, monitoring the health of the people remotely using smart clothes with sensors that can monitor their vital signs, and keeping track of fleets of cars and buses. The variety of possible applications that can result in M2M communications is very large and applicable to almost any industry one can imagine.

M2M communication has a different set of characteristics than communication between humans and a computer. The communication between humans is heavily dominated by audio (e.g., voice communication or music downloads), video (e.g., movies downloaded or streamed from the Internet), and web surfing using the HTTP protocol. M2M communication will rarely use audio or video content and can use communication protocols that are far more efficient than HTTP. The M2M devices communicate using short amounts of data and the communication would tend to be highly predictable, for example, a car or meter will report its reading at a regular period.

M2M uses the same mobile data network that is used for mobile phones to communicate. Due to the differences in the pattern of communications, they may require a different architectural model to support them from the mobile network operators. However, M2M communications do not add in any significant manner to the data being exchanged on the network.

For a more detailed discussion of M2M, please refer to Reference 71.

14.2 INTERNET OF THINGS

Internet of Things is closely related to M2M and represents a state where the Internet connects not just servers, computers, and mobile devices but everyday devices as well. The devices the Internet of Things connects can include coffee mugs, refrigerators, automobile parts, microwaves, and household appliances. One can imagine a variety of applications from Internet of Things, for example, smart refrigerators that will place proactive orders for groceries so that one never runs out of supplies; dolls that are connected to speech processing software over the network so they can carry an intelligent conversation with a child, houses that anticipate your predicted time of arrival to the home and adjust the thermostat and other comfort controls appropriately. The Internet of Things would provide an overall network which will be a superset of M2M, mobile phone-based connectivity, and other types of sensor networks which today are not connected in a global Internet.

There are several variations and architectures proposed for Internet of Things [72,73], and they usually deal with issues such as heterogeneity of different devices, scalability in addressing the large number of devices that will be introduced by Internet of Things, which will be a few orders of magnitude larger than the current Internet, and scalability in processing and collecting information from such a large set of devices in detail.

Internet of Things, when it is attained, will also cause an increase in the amount of data that is the subject of this book. However, Internet of Things also has several orthogonal issues which need to be addressed, for example, scalability, federation across multiple devices, and interoperability among diverse schemes for everyday things to work in a seamless manner.

14.3 PARTICIPATORY SENSING

A concept related to the Internet of Things is that of participatory sensing. Many computing systems that are deployed today have the task of collecting information from various sources. Participatory sensing is the idea [74,75] that a significant amount of data collection can be done by the mobile phones that people carry around with them. With the large number of cell phones that are ubiquitously present in modern society, information collected by the cell phones can provide a sensing infrastructure that will allow collection of real-time information at an unprecedented scale.

The nature and type of information collected by participatory sensing depends on the sensing capabilities of the Smartphones that are participating. Assuming that the sensors are just photographs captured by people and uploaded by users to a central location, participatory sensing can provide insights into the current snapshot of a city, for example, any potholes in the roads that may need repair or information about the vegetation of a city and how it changes over time. If cell phones are equipped with sensors that can measure air quality, they can keep track of pollution in the city and allow better tracking of natural resources.

Participatory sensing refers to the use of information from mobile phones that people are carrying with them. However, nonparticipatory use of Smartphones for sensing is also available as an instance of an Internet of Things or M2M application. If one needs to make up a quick infrastructure for collecting information over a wide area, for example, if you wanted to create a system that tracked the amount of foot traffic that is occurring near each of the four potential locations you may have for opening a retail store, you can use mobile phones in lieu of customized cameras for information collection. Obtaining cameras, hooking them together into a network, and creating the system that counted the number of people passing through could take a substantial amount of time in getting the infrastructure installed. Obtaining a few phones and enabling them to record and collect information into a central location would be a much faster way to get the desired system operational.

14.4   MOBILE TRANSFORMATION OF BUSINESS

Mobile devices have the potential to cause a disruptive transformation in the manner by which business is conducted in many businesses and enterprises. To a large extent, this transformative impact of mobile on business is what has caused an explosion in mobile data, the subject of this book.

Any type of IT transaction that happens in the commercial arena requires an infrastructure consisting of three components, a client, a server, and a network connecting the clients to the servers. In most of the cases, the server will be located at a data center owned by the commercial enterprise, the network would typically be a private network operated by the enterprise. The client is different for different types of transactions. At a retail store, the client is the point of sale terminal. At any branch of a bank, the client is a computer operated by the teller or the automated teller machine. For a delivery company, the client is a scanning device which the delivery person uses to scan tracking information at every delivery, and the network could be a cellular or satellite connection. When one pays fare on a cab using a credit card, the client is a credit card reader connected via a cellular network.

Each different type of client device which is used in the business operation has an associated cost of owning, operating, and maintaining it. For a large number of such devices, the capabilities of these devices can be provided by a Smartphone. Just like the use of Smartphones as sensors in Section 14.3, this case is the use of Smartphones for other operations that are currently done by specialized clients. Because of the reduction in cost that can be achieved by switching from specialized clients to mobile phones, we would envision a future in which applications on mobile phones replace the role of clients in many of the business IT transactions. Depending on the nature of the business, the mobile application may be running on a mobile device belonging to a customer of the enterprise or it may be running on a mobile device belonging to an employee of the enterprise.

This has the potential for changing the way many businesses operate. A mobile application on a customer mobile device can be used to pay for purchases at a retail store, with just a security check being made as the customer leaves the retail store. Taxi cabs could be paid using digital cash applications on the mobile devices of the riders. Several banks have already started allowing several types of banking transactions using mobile applications on their customer devices. The tracking of packages for any delivery company can eliminate special scanners and trackers, using mobile applications for the same purpose. Restaurants can make reservations for their customers, as well as order their supplies and track orders using mobile device-based applications. There is virtually no business or industry that will be left untouched by the replacement of specialized clients with a mobile device-based application.

As mobile applications get more widely deployed within the business arena, the techniques described in this book will become more and more relevant to creating a infrastructure that can sustain the mobile business with sufficient quality of service.

14.5   SOFTWARE-DEFINED NETWORKS

SDN is a new approach to develop networking protocols, in which an explicit separation is made between the control aspects of networking and the data transfer aspects of networking. In connection-oriented networks, the set of protocols which were used to establish the connections comprised the set of control protocols. In contrast, the data protocols were the ones used to transfer the information among users of the network. In telephony, the set of exchanges which happened to enable a person to dial another number and have the phone ring are done using the control protocols while the actual conversation flows on data protocol.

While the separation of network control and data is very explicit in communication protocols that have evolved with a legacy of telephone companies, for example, most of the cellular protocols, the separation between control and data is relatively ill defined in the TCP/IP family of protocols. When voice over IP emergence led to the development of the Session Initiation Protocol (SIP), it was the first explicit control protocol in this protocol family. Nevertheless, IP needed a set of routing protocols to determine how packets will be routed in the network. Similarly, the most common MAC protocol in the family, namely Ethernet, needed a set of exchanges to determine a spanning tree among different interconnected Ethernet switches so that they could communicate to each other. These exchanges can be viewed as an implicit control protocol for the TCP/IP protocol family.

SDN [76] separates out the control protocols for IP and Ethernet in an explicit manner and allows the control operations to be implemented via software in a central location, called the controller. A standard protocol OpenFlow [77] has been defined which allows the controller to determine the right forwarding behavior of packets arriving at the switches/routers in the network. The switches in themselves can become simpler by eliminating the bulk of control protocol processing. The controller would allow the incorporation of new techniques for forwarding packets by writing software-based extensions. The behavior of the entire network can thus be controlled by means of software applications written in a central location.

SDN can be used to modify network behavior rapidly. As an example, standard Ethernet switches control organizes all the switches into a spanning tree and packets can only be forwarded along the spanning tree. This may not always be the best route in all cases. Using a centralized controller, the switches forwarding tables can be modified to use alternate paths as needed, thereby enabling new networking routing/switching behaviors more rapidly than trying to develop an alternative distributed approach.

While SDN is not directly related to subject of mobile data growth, one of its benefits is the ability to reduce the costs of operating a network. The cost reduction comes as an instance of consolidation, where complexity in distributed network elements in reduced and a central software system takes on that onus. In that respect, it can be viewed as one of the cost-reduction mechanisms that many network operators may like to use.

14.6   MOBILE FIRST PHILOSOPHY

The mobile first philosophy refers to the concept that applications and services ought to be developed with mobile device access as a primary user in mind. This is in contrast to a model in which mobile device access is added as an afterthought, with the primary focus being web-based usage. As discussed in Chapter 13, many enterprises have existing applications and services that were designed for access from laptops and personal computers over the web, and there are various ways to adapt existing application for better user experience on mobile devices.

Mobile devices are seeing a much bigger growth in volume than traditional computers or laptops. This means that the primary set of users for any service would be from mobile systems and mobile users. Mobile first philosophy calls for developers of Internet-based services to take the account of these users as the primary input into the development of the services. This is in contrast to a "mobile second" approach where services are designed for web-based users and mobile interfaces are grafted on as an afterthought.

The mobile first philosophy requires incorporating the needs of the mobile users when designing the user interface, content layout, and protocol exchanges of a web site or Internet-accessible service. It also includes giving proper attention to issues of managing mobile applications, maintaining them, and developing them across multiple platforms.

As mobile devices continue to surge ahead in their popularity, the philosophy of mobile first is likely to become the common guiding principle under which different applications are developed.

14.7   NETWORK ANALYTICS

During the course of operation of a network, a tremendous amount of data is generated. By processing that data, a significant amount of insights into the behavior of the network and its users can be obtained. The insights obtained from the processing of the information can be used in a variety of ways, for example, modify the configuration of the network, offer new services, and improve customer service. Some of the approaches for creating new data monetization services and bandwidth reduction can be viewed as applications of network analytics.

Network analytics is the area of processing and analyzing the information generated within the network and feed it into a decision-making system. Data that is generated into the network can be broadly considered to fall into two categories: data in motion and data at rest. Data in motion is data that is not stored or stored only for a limited period of time. Data at rest is data that is stored in a stable repository and expected to remain in that repository as long as required. An example of data in motion is the contents of network packets flowing through a mobile network. At rates exceeding several hundreds of gigabits per second, the contents of the packets flowing through a network cannot be stored for historical analysis. Any processing that needs to be done on the packets will need to be done in real time. An example of data at rest is the set of call data records that are generated for people making telephone calls on the network. The call data records are used for billing customers and can be archived for multiple years. The amount of time a specific type of data is stored in the network depends on the retention policies of the network operator.

Network analytics for data at rest is essentially applying the techniques of data mining [78,79] to data collected from the network. The specific technique that can be used for analysis depends on the use-case driving network analytics on the system. Network analytics for data in motion needs to use algorithms that can be performed in a limited budget of time. They are applications of stream processing [80] to the domain of networks. In essence, stream processing performs fast analysis on the information streaming through the network while data mining performs more thorough processing of the information that is stored historically.

In stream processing, or network analytics on data in motion, an important concept is that of predictive analytics. Predictive analytics use current information to predict the state of the network some time in the future. As an example, information about the traffic on a link can be used to predict the expected traffic a few minutes in the future. Prediction allows the network to react better to anticipated conditions in the near future, since any reconfiguration or modification of the network would have an operational latency.

Network analytics is an immensely valuable tool for any type of network. Its applications include issues related to mobile data growth, the topic of this book. However, its scope is much bigger than the topic of the book alone, and it has many applications in various aspects of network operations and management, from network management to customer service and formulation of billing plans.

14.8 CONCLUSIONS

Mobile computing and mobile applications have the potential to change the way business is conducted and may have a substantial impact on the daily lives of people as the vision of Internet of Things is achieved. We are at an interesting time in human history where new advances in mobile computing have the potential to transform the way we live, play, work, and conduct business. The replacement of specialized clients for information collection and conducting business can cause a significant increase in amount of data traffic. As the data traffic increases, the techniques described in this book will become more relevant and can be applied in many different ways.

REFERENCES

[1] Cisco Systems. (2010). Cisco Visual Networking Index: Global Mobile Data Traffic Forecast Update 2009–2014, http://www.cisco.com/en/US/solutions/collateral/ns341/ns525/ns537/ns705/ns827/white_paper_c11-520862.html, last retrieved November 8, 2013. Cisco Public Information, February 9.

[2] Sandvine. (2011). Global Internet Phenomena Report, https://www.sandvine.com/trends/global-internet-phenomena/, last retrieved November 8, 2013.

[3] Meeker, M. (2012). Internet Trends 2012, http://www.kpcb.com/file/kpcb-internet-trends-2012, last retrieved March 3, 2013.

[4] The Cooperative Association for Internet Data Analysis, CAIDA's Annual Report for 2012, http://www.caida.org/home/about/annualreports/2012/, last retrieved November 3, 2013.

[5] Williamson, C. (2001). Internet traffic measurement. IEEE Internet Computing, 5(6), 70–74.

[6] ByteMobile. (2013). Mobile Analytics Report, February 2013, http://www.bytemobile.com/news-events/mobile_analytics_report.html, last retrieved March 3, 2013.

[7] Nygren, E., Sitaraman, R. K., & Sun, J. (2010). The Akamai network: a platform for high-performance Internet applications. ACM SIGOPS Operating Systems Review, 44(3), 2–19.

[8] Maier, G., Feldmann, A., Paxson, V., & Allman, M. (2009). On dominant characteristics of residential broadband internet traffic. In Proceedings of the

Techniques for Surviving the Mobile Data Explosion, First Edition.
Dinesh Chandra Verma and Paridhi Verma.
© 2014 The Institute of Electrical and Electronics Engineers, Inc. Published 2014 by John Wiley & Sons, Inc.

9th ACM SIGCOMM conference on Internet measurement conference, Barcelona, August 17–21. ACM, New York, pp. 90–102.

[9] Wang, J. (1999). A survey of web caching schemes for the internet. ACM SIGCOMM Computer Communication Review, 29(5), 36–46.

[10] Fielding, R., Gettys, J., Mogul, J., Frystyk, H., Masinter, L., Leach, P., & Berners-Lee, T. (1999). Hypertext Transfer Protocol—HTTP/1.1, http://www. ietf.org/rfc/rfc2616.txt, last retrieved November 8, 2013.

[11] Verma, D. C. (2002). Content Distribution Networks: An Engineering Approach, Wiley-Interscience, New York.

[12] Ziv, J., & Lempel, A. (1977). A universal algorithm for sequential data compression. IEEE Transactions on Information Theory, 23(3), 337–343.

[13] Ziv J., & Lempel A. (1978). Compression of individual sequences via variable-rate coding. IEEE Transactions on Information Theory, 24(5), 530–536.

[14] Salomon, D. (2004). Data Compression: The Complete Reference, Springer-Verlag Incorporated, New York.

[15] Sayood, K. (2005). Introduction to Data Compression, Morgan Kaufmann, Amsterdam, the Netherlands.

[16] BlueCoat Systems. (2007). Blue Coat Systems. Technology Primer: Byte Caching, http://www.bluecoat.com/sites/default/files/documents/files/ Byte_Caching.a.pdf, last retrieved March 3, 2013.

[17] Hewlett Packard Company. (1995). HP Case Study: WAN Link Compression on HP Routers, http://www.hp.com/rnd/support/manuals/pdf/comp.pdf, last retrieved November 8, 2013.

[18] Cisco. (2006). WAN Compression FAQs, Document No. 9289, http://www.cisco. com/image/gif/paws/9289/wan_compression_faq.pdf, last retrieved November 8, 2013.

[19] Jacobson, V. (1990). RFC 1144: Compressing TCP/IP headers for low-speed serial links, http://www.rfc-editor.org/rfc/rfc1144.txt, last retrieved November 8, 2013.

[20] Degermark, M., Nordgren, B., Pink, S., & Compression, I. H. (1999). RFC 2507: IP Header Compression, http://www.rfc-editor.org/rfc/rfc2507.txt, last retrieved November 8, 2013.

[21] Engan, M., Casner, S., & Bormann-RFC, C. (1999). RFC 2509: Compressing IP headers for PPP, http://www.rfc-editor.org/rfc/rfc2509.txt, last retrieved November 8, 2013.

[22] Tye, C. S., & Fairhurst, G. (2003). A review of IP packet compression techniques. In Proceedings of PostGraduate networking conference, June, Liverpool, UK. Liverpool John Moores University, Liverpool, UK.

[23] Quinn, B., & Almeroth, K. RFC 3170: IP Multicast Applications: Challenges and Solutions, http://www.rfc-editor.org/rfc/rfc3170.txt, last retrieved November 8, 2013.

[24] Kernen, T., & Simlo, S. (2010). AMT – Automatic IP multicast without explicit tunnels, European Broadcasting Union Technical Review, Q4.

[25] Chennikara, J., Chen, W., Dutta, A., & Altintas, O. (2002). Application-layer multicast for mobile users in diverse networks. In IEEE Global telecommunications conference, November 17–21. IEEE, Taipei, Taiwan.

[26] Pancha, P., & El Zarki, M. (1994). MPEG coding for variable bit rate video transmission. IEEE Communications Magazine, 32(5), 54–66.

[27] Li, W. (2001). Overview of fine granularity scalability in MPEG-4 video standard. IEEE Transactions on Circuits and Systems for Video Technology, 11(3), 301–311.

[28] Vandalore, B., Feng, W. C., Jain, R., & Fahmy, S. (2001). A survey of application layer techniques for adaptive streaming of multimedia. Real-Time Imaging, 7(3), 221–235.

[29] China Mobile Research Institute. (2010). C-RAN: The Road Toward Green RAN, White Paper, http://ss-mcsp.riit.tsinghua.edu.cn/cran/C-RAN%20China COM-2012-Aug-v4.pdf, last retrieved March 3, 2013.

[30] Lin, Y., Shao, L., Zhu, Z., Wang, Q., & Sabhikhi, R. K. (2010). Wireless network cloud: architecture and system requirements. IBM Journal of Research and Development, 54(1), 1–12.

[31] Dillinger, M., Madani, K., & Alonistioti, N. (2003). Software Defined Radio: Architectures, Systems, and Functions, John Wiley & Sons, Inc, Hoboken, NJ, pp. 191–206.

[32] Lin, Y. D., & Hsu, Y. C. (2000). Multihop cellular: a new architecture for wireless communications. In Proceedings of IEEE INFOCOM, March 26–30, Tel Aviv, Israel. IEEE, Taipei, Taiwan, pp. 1273–1282.

[33] Chandrasekhar, V., Andrews, J., & Gatherer, A. (2008). Femtocell networks: a survey. IEEE Communications Magazine, 46(9), 59–67.

[34] IBM & Nokia Siemens Networks. (2013). IBM and Nokia Siemens Networks Announce World's First Mobile Edge Computing Platform, http://www-03.ibm.com/press/us/en/pressrelease/40490.wss, last retrieved March 3, 2013.

[35] British Telecom Web Article. (2012). Eco-Friendly 'Network Virtualisation' Saves Money, http://www.btplc.com/Innovation/News/NetworkVirtualization.htm, last retrieved July 10, 2012.

[36] Neuman, B., & Ts'o, T. (1994). Kerberos: an authentication service for computer networks. IEEE Communications Magazine, 32(9), 33–38.

[37] OpenID Foundation. (2010). An Introduction to OpenID and the OpenID Foundation 2011, http://openid.net/wordpress-content/uploads/2011/03/Introduction-to-OpenID-Foundation-March-2011.pdf, last retrieved August 20, 2012.

[38] Wang, R., Chen, S., & Wang, X. (2012). Signing me onto your accounts through facebook and google: a traffic-guided security study of commercially deployed single-sign-on web services. In IEEE Symposium on Security and Privacy 2012, May 20–23, San Francisco, CA. IEEE, Taipei, Taiwan, pp. 365–379.

[39] Dingledine, R., Mathewson, N., & Syverson, P. (2004). Tor: the second-generation onion router. In Proceedings of the 13th USENIX Security Symposium, August 9–13, San Francisco, CA. USENIX Association, Berkeley, CA.

[40] Hughes, N., & Lonie, S. (2007). M-PESA: mobile money for the "unbanked" turning cellphones into 24-hour tellers in Kenya. Innovations: Technology, Governance, Globalization, 2(1–2), 63–81.

[41] Nokia Siemens Networks. (2009). The Impact of Latency on Application Performance, http://www.nokiasiemensnetworks.com/system/files/document/ LatencyWhitepaper.pdf, last retrieved March 1, 2013.

[42] Bonomi, F., Milito, R., Zhu, J., & Addepalli, S. (2012). Fog computing and its role in the internet of things. In Proceedings of the first edition of the MCC workshop on mobile cloud computing, August 17, Helsinki, Finland. ACM, New York, pp. 13–16.

[43] Satyanarayanan, M., Bahl, P., Caceres, R., & Davies, N. (2009). The case for VM-based cloudlets in mobile computing. IEEE Pervasive Computing, 8(4), 14–23.

[44] Padmanabhan, V. N., & Subramanian, L. (2001). An investigation of geographic mapping techniques for internet hosts. ACM SIGCOMM Computer Communication Review, 31(4), 173–185.

[45] Siwpersad, S., Gueye, B., & Uhlig, S. (2008). Assessing the geographic resolution of exhaustive tabulation for geolocating Internet hosts. Passive and Active Network Measurement, Springer-Verlag, Berlin/Heidelberg, Germany, pp. 11–20.

[46] Mobile Operating Systems. (2013). http://en.wikipedia.org/wiki/Mobile_ operating_system, last retrieved July 30, 2013.

[47] Android Version History. (2013). http://en.wikipedia.org/wiki/Android_ version_history, last retrieved July 30, 2013.

[48] Jones, C. (2009). Software Engineering Best Practices: Lessons from Successful Projects in the Top Companies, McGraw-Hill Osborne, Emeryville, CA.

[49] Sommerville, I. (2010). Software Engineering, Addison-Wesley, Harlow, UK.

[50] Welch, G. F. (1995). A survey of power management techniques in mobile computing operating systems. ACM SIGOPS Operating Systems Review, 29(4), 47–56.

[51] Vallina-Rodriguez, N., & Crowcroft, J. (2012). Energy management techniques in modern mobile handsets. IEEE Communications Surveys and Tutorials, 19, 1–20.

[52] Abderazek, B. A., & Sowa, M. (2007). Advanced power management techniques for mobile communication systems. In I. K. Ibrahim and D. Taniar (eds.), Mobile Multimedia: Communication Engineering Perspective. Nova Publishers, New York, pp. 259–278.

[53] Carroll, A., & Heiser, G. (2010). An analysis of power consumption in a smartphone. In Proceedings of the 2010 USENIX conference on USENIX annual technical conference, June 22–25, Boston, MA. USENIX Association, Berkeley, CA, pp. 21–21.

[54] Vallina-Rodriguez, N., Hui, P., Crowcroft, J., & Rice, A. 2010. Exhausting battery statistics: understanding the energy demands on mobile handsets. In ACM MobiHeld, August 30–Septemeber 3, New Delhi, India. ACM, New York, pp. 9–14.

[55] Pathak, A., Hu, Y. C., & Zhang, M. (2012). Where is the energy spent inside my app?: fine grained energy accounting on smartphones with eprof. In Proceedings of the 7th ACM European conference on Computer Systems, April 14–17, Prague, Czech Republic. ACM, New York, pp. 29–42.

[56] Frankk, D. (2013). Best Practices for Mobile Application Development, http://ezinearticles.com/?Best-Practices-for-Mobile-Application-Development&id=6262034, last retrieved February 17, 2013.

[57] Mikhalenko, P. (2013). Best Practices for Mobile Web Application Development, http://www.techrepublic.com/article/best-practices-for-mobile-web-application-development/6095452, last retrieved February 17, 2013.

[58] Connors, A., & Sullivan, B. (2012). W3C Recommendations – Mobile Web Application Best Practices, http://www.w3.org/TR/mwabp/, last retrieved November 8, 2013.

[59] Chen, G., & Kotz, D. (2000). A Survey of Context-Aware Mobile Computing Research (Vol. 1, No. 2.1, pp. 1–15). Technical Report TR2000-381, Department of Computer Science, Dartmouth College, Hanover, NH.

[60] Scott, J., Crowcroft, J., Hui, P., & Diot, C. (2006). Haggle: a networking architecture designed around mobile users. In WONS 2006: third annual conference on wireless on-demand network systems and services, January 18–20, Les Menuires, France. International Federation for Information Processing, Laxenburg, Austria, pp. 78–86.

[61] Almeida Bittencourt, R., & Carr, D. A. (2001). A method for asynchronous, web-based lecture delivery. In Frontiers in education conference, 2001. 31st Annual (Vol. 2, pp. F2F-12), October 10–13, Reno, NV. IEEE, Taiwan, China, pp. 12–17.

[62] Marmasse, N., & Schmandt, C. (2000). Location-aware information delivery with ComMotion. In Proceedings of second international symposium on Handheld and Ubiquitous Computing, September 25–27, Bristol. Springer, London, pp. 157–171.

[63] Qian, F., Wang, Z., Gerber, A., Mao, Z., Sen, S., & Spatscheck, O. (2011). Profiling resource usage for mobile applications: a cross-layer approach. In Proceedings of the 9th international conference on mobile systems, applications, and services, June 28–July 1, Washington, DC. ACM, New York, pp. 321–334.

[64] Felt, A. P., Finifter, M., Chin, E., Hanna, S., & Wagner, D. (2011). A survey of mobile malware in the wild. In Proceedings of the 1st ACM workshop on security and privacy in smartphones and mobile devices, October 17–21, Chicago, IL. ACM, New York, pp. 3–14.

[65] Shabtai, A., Fledel, Y., Kanonov, U., Elovici, Y., & Dolev, S. (2009). Google Android: a state-of-the-art review of security mechanisms. IEEE Security & Privacy, 8(2), 35–44.

[66] Rosenblatt, S. (2013). Don't get Faked by Android Antivirus Apps, http://download.cnet.com/8301-2007_4-57391170-12/dont-get-faked-by-android-antivirus-apps/, last retrieved February 28, 2013.

[67] Krawczyk, H. (1994). Secret sharing made short. In Advances in cryptology—CRYPTO'93, Springer, Berlin/Heidelberg, pp. 136–146.

[68] Tilley, S., Toeter, B., & Wong, K. (2001). Issues in accessing web sites from mobile devices. In Proceedings of IEEE workshop on web site evolution, November 10, Florence, Italy. IEEE, Taiwan, China, pp. 97–104.

[69] Trewin, S. (2006). Physical usability and the mobile web. In Proceedings of ACM international cross-disciplinary workshop on Web accessibility, May 22–23, Edinburgh, Scotland. ACM, New York, pp. 109–112.

[70] Brandenburg, M. (2010). Mobile Enterprise Application Platforms: A Primer, http://searchconsumerization.techtarget.com/tutorial/Mobile-enterprise-application-platforms-A-primer, last retrieved February 28, 2013.

[71] Boswarthick, D., Elloumi, O., & Hersent, O. (2012). M2M Communications: A Systems Approach, John Wiley & Sons, Inc, Hoboken, NJ.

[72] Boswarthick, D., Elloumi, O., & Hersent, O. (2012). The Internet of Things: Key Applications and Protocols, John Wiley & Sons, Ltd, Chichester, UK.

[73] Atzori, L., Iera, A., & Morabito, G. (2010). The internet of things: a survey. Computer Networks, 54(15), 2787–2805.

[74] Burke, J. A., Estrin, D., Hansen, M., Parker, A., Ramanathan, N., Reddy, S., & Srivastava, M. B. (2006). Participatory Sensing, http://escholarship.org/uc/item/19h777qd, last retrieved February 28, 2013.

[75] Estrin, D. L. (2010). Participatory sensing: applications and architecture. In Proceedings of the 8th international conference on mobile systems, applications, and services, June 15–18, San Francisco, CA. ACM, New York, pp. 3–4.

[76] Open Networking Foundation. (2013). SDN Definition, https://www.open networking.org/sdn-resources/sdn-definition, last retrieved July 30, 2013.

[77] Mckeown, N., Anderson, T., Balakrishnan, H., Parulkar, G., Peterson, L., Rexford, J., Shenker, S., & Turner, J.(2008). OpenFlow: enabling innovation in campus networks. ACM SIGCOMM Computer Communication Review. 38(2), 69–74.

[78] Kantardzic, M. (2002). Data Mining: Concepts, Models, Methods, and Algorithms, IEEE Press, Hoboken, NJ.

[79] Han, J., Kamber, M., & Pei, J. (2011). Data Mining: Concepts and Techniques. Morgan Kaufman, Waltham, MA.

[80] Salehi, A. (2010). Low Latency, High Performance Data Stream Processing: Systems Architecture, Algorithms and Implementation, VDM Verlag, Saarbrücken, Germany.

INDEX

Techniques for Surviving the Mobile Data Explosion, First Edition.
Dinesh Chandra Verma and Paridhi Verma.
© 2014 The Institute of Electrical and Electronics Engineers, Inc. Published 2014 by John Wiley & Sons, Inc.

IEEE PRESS SERIES ON
DIGITAL AND MOBILE COMMUNICATION

John B. Anderson, *Series Editor*
University of Lund

CPSIA information can be obtained at www.ICGtesting.com
Printed in the USA
BVOW01n1813260314

348867BV00003B/3/P